STEM CELLS – LABORATORY AND CLINICAL RESEARCH SERIES

NEURAL STEM CELLS AND CELLULAR THERAPY

Stem Cells - Laboratory and Clinical Research Series

Focus on Stem Cell Research
Erik V. Greer (Editor)
2004. ISBN: 1-59454-043-8

Trends in Stem Cell Research
Erik V. Greer (Editor)
2005. ISBN: 1-59454-315-1

New Developments in Stem Cell Research
Erik V. Greer (Editor)
2006. ISBN: 1-59454-847-1

Neural Stem Cell Research
Erik V. Greer (Editor)
2006. ISBN: 1-59454-846-3

Stem Cell Therapy
Erik V. Greer (Editor)
2006. ISBN: 1-59454-848-X

Embryonic Stem Cell Research
Erik V. Greer (Editor)
2006. ISBN: 1-59454-849-8

Frontiers in Stem Cell Research
Julia M. Spanning (Editor)
2006. ISBN: 1-60021-294-8

Stem Cells and Cancer
Devon W. Parsons (Editor)
2007. ISBN: 1-60021-517-3

Hematopoietic Stem Cell Transplantation Research Advances
Karl B. Neumann (Editor)
2008. ISBN: 978-1-60456-042-8

Stem Cell Applications in Diseases
Mikkel L. Sorensen (Editor)
2008. ISBN: 978-1-60456-241-5

Stem Cell Applications in Diseases
Mikkel L. Sorensen (Editor)
2008. ISBN: 978-1-60876-925-4
(Online Book)

Leading-Edge Stem Cell Research
Prasad S. Koka (Editor)
2008. ISBN: 978-1-60456-268-2

Stem Cell Research Progress
Prasad S. Koka (Editor)
2008. ISBN: 978-1-60456-308-5

Stem Cell Research Progress
Prasad S. Koka (Editor)
2008. ISBN: 978-1-60876-924-7
(Online Book)

Progress in Stem Cell Applications
Allen V. Faraday and Jonathon T. Dyer (Editors)
2008. ISBN: 978-1-60456-316-0

Developments in Stem Cell Research
Prasad S. Koka (Editor)
2008. ISBN: 978-1-60456-341-2

Developments in Stem Cell Research
Prasad S. Koka (Editor)
2008. ISBN: 978-1-60741-213-7
(Online Book)

Stem Cells and Regenerative Medicine. Volume 1. Adult Neurogenesis and Neural Stem Cells
Philippe Taupin
2008. ISBN: 978-1-60456-472-3

Stem Cells and Regenerative Medicine. Volume 2. Embryonic and Adult Stem Cells
Philippe Taupin
2008. ISBN: 978-1-60456-473-0

Stem Cells and Regenerative Medicine. Volume 3. Pharmacology and Therapy
Philippe Taupin
2008. ISBN: 978-1-60456-474-7

Stem Cells and Regenerative Medicine. Volume 4. Neurological Diseases and Cellular Therapy
Philippe Taupin
2010. ISBN: 978-1-60741-783-5

Stem Cells and Regenerative Medicine. Volume 5. Patents and Clinical Trials
Philippe Taupin
2010. ISBN: 978-1-60741-782-8

Gut Stem Cells: Multipotent, Clonogenic and the Origin of Gastrointestinal Cancer
Shigeki Bamba and William R. Otto
2008. ISBN: 978-1-60456-968-1

Stem Cell Transplantation, Tissue Engineering and Cancer Applications
Bernard N. Kennedy (Editor)
2008. ISBN: 978-1-60692-107-4

Stem Cells
Philippe Taupin
2009. ISBN: 978-1-60692-214-9

Stem Cell Plasticity
Suraksha Agrawal, Piyush Tripathi and Sita Naik
2009. ISBN: 978-1-60741-473-5

Neural Stem Cells and Cellular Therapy
Philippe Taupin
2010. ISBN: 978-1-60876-017-6

STEM CELLS – LABORATORY AND CLINICAL RESEARCH SERIES

NEURAL STEM CELLS AND CELLULAR THERAPY

PHILIPPE TAUPIN

Nova Biomedical Books
New York

For permission to use material from this book please contact us:
Telephone 631-231-7269; Fax 631-231-8175
Web Site: http://www.novapublishers.com

Library of Congress Cataloging-in-Publication Data

Taupin, Philippe.
Neural stem cells and cellular therapy / Philippe Taupin.
p. ; cm.
Includes bibliographical references and index.
ISBN 978-1-60876-017-6 (hardcover)
1. Neural stem cells. 2. Developmental neurobiology. I. Title.
[DNLM: 1. Adult Stem Cells--physiology. 2. Neurons--physiology. 3. Central Nervous System Diseases--therapy. 4. Nerve Regeneration. 5. Neurogenesis. 6. Stem Cell Transplantation. WL 102.5 T227n 2009] QP363.5.T392 2009
612.6'4018--dc22 2009035724

Published by Nova Science Publishers, Inc. ✢ New York

Contents

Introduction
Adult Neural Stem Cells: The Promise of the Future

Abstract

Stem cells are self-renewing undifferentiated cells that give rise to multiple types of specialized cells of the body. In the adult, stem cells are multipotents and contribute to homeostasis of the tissues and regeneration after injury. Until recently, it was believed that the adult brain was devoid of stem cells, hence unable to make new neurons and regenerate. With the recent evidences that neurogenesis occurs in the adult brain and neural stem cells (NSCs) reside in the adult central nervous system (CNS), the adult brain has the potential to regenerate and may be amenable to repair. The function(s) of NSCs in the adult CNS remains the source of intense research and debates. The promise of the future of adult NSCs is to redefine the functioning and physiopathology of the CNS, as well as to treat a broad range of CNS diseases and injuries.

Introduction

Seminal studies in the 60s using [3H]-thymidine autoradiographic labeling by Altman and Das were the first to report the generation of new neuronal cells in the adult rodent dentate gyrus (DG), and cell proliferation in the ventricular zone, migration and persisting neurogenesis in the adult olfactory bulb (OB) (Altman and Das, 1965; Altman, 1969). However, these studies had little impact, because of the paucity of cells labeled and the difficulty of definitively identifying them. It was not until the 1990s, with the advent of new procedures for labeling dividing cells in the CNS, like bromodeoxyuridine (BrdU) (Gratzner, 1982; Miller and Nowakowski, 1988) and retroviral labelings (Van Praag et al., 2002), that neurogenesis in the SVZ and DG really became accepted (Gross, 2000; Taupin and Gage, 2002). Though over the past decades, significant progresses have been made in the field of adult neurogenesis and NSCs (table 1), there is much debate, controversies and questions to be answered.

Adult Neurogenesis, Facts and Debates

Neurogenesis in the Adult Mammalian Brain

Neurogenesis occurs primarily in two areas of the adult brain in mammals: the DG of the hippocampus and the subventricular zone (SVZ) in several species, including human (Eriksson et al., 1998; Curtis et al., 2007a). In the DG, newly generated neuronal cells in the subgranular zone (SGZ) migrate to the granular layer, where they differentiate into mature neuronal cells, and extend axonal projections to the CA3 area in rodents and primates. In the SVZ, cells are generated in the anterior part of the SVZ and migrate to the OB, through the rostro-migratory stream (RMS), where they differentiate into interneurons of the OB in rodent and non-human primates (Taupin P and Gage, FH. 2002). Newly generated neuronal cells establish functional connections with neighboring cells (Van Praag et al., 2002; Carlen et al., 2002), particularly GABAergic innervations in the DG, soon after their migration is completed (Wang et al., 2005). As many as 9,000 new neuronal cells - or 0.1% of the granule cell population - are generated per day in the DG of mice, and 65-75% of the bulbar neurons are replaced during a 6 weeks period in young adult rats (Kempermann et al., 1997; Kato et al., 2001; Cameron and McKay, 2001). Among them, a significant proportion undergoes programmed cell death rather than achieving maturity (Morshead and van der Kooy, 1992; Cameron and McKay, 2001; Gould et al., 2001).

The newly generated neuronal cells that survived to maturity may be very stable, and may permanently replace cells born during development, as adult-generated neuronal cells have been reported to survive for extended period of time (e.g. for at least 2 years in human DG) (Altman and Das, 1965; Eriksson et al., 1998; Dayer et al., 2003; Kempermann et al., 2003). Neurogenesis may also occur, albeit at lower levels, in other areas of the mammalian brain, like the Ammon's horn CA1, neocortex and substantia nigra (SN) (Gould et al., 1999; Rietze et al., 2000; Zhao et al., 2003). However, some of these reports have been contradicted by other studies (Kornack and Rakic, 2001; Lie et al., 2002; Frielingsdorf et al., 2004; Gould, 2007). Hence, the bulk of evidence suggests that there is little if any neurogenesis going on constitutively in other brain regions.

Stem Cells in the Adult Brain

The origin of newly generated neuronal cells in the adult brain remains the source of controversies. One theory contends that they originate from differentiated ependymal cells in the lateral ventricle, while another contends that they originate from astrocyte-like cells in the SVZ and SGZ (Taupin and Gage, 2002). A glial origin for adult generated neuronal cell receives further support recently (Filippov et al., 2003; Garcia et al., 2004). Hence, the possibility of ependymal origins for NSCs has been has mostly discounted and astrocyte-like cells represent the most accepted model for the source of stem cells of the adult brain.

It is postulated that newly generated neuronal cells originate from residual stem cells in the adult brain. Stem cells are defined by five attributes: proliferation, self-renewal over an extended period of time, generation of a large number of differentiated progeny, maintenance of the homeostasis of the tissue and regeneration of the tissue following injury (Potten and Loeffler, 1990). NSCs are the self-renewing, multipotent cells that generate neurons, astrocytes and oligodendrocytes of the nervous system. Neural progenitor cells are, as most broadly defined, any cells that do not fulfill all of the attributes of NSCs. Though NSCs remain to be characterized in the adult CNS, self-renewing, multipotent NSC-like cells have been isolated and characterized in vitro from various areas of the adult CNS, neurogenic and non-neurogenic, including the spinal cord, suggesting that NSC may reside throughout the CNS (Taupin and Gage, 2002).

There are currently no specific markers of adult NSCs. Nestin, the transcription factors sox-2, oct-3/4 and the RNA binding protein Musashi 1 are markers for neural progenitor and stem cells, but also label population of glial cells (Lendahl et al., 1990; Sakakibara et al., 1996; Doetsch et al., 1999; Zappone et al., 2000; Kaneko et al., 2000; Komitova et al., 2004; Okuda et al., 2004), further fueling the controversies over the origin of newly generated neuronal cells in the adult brain.

Rate and Modulation

The rate of neurogenesis in the rodent DG and SVZ is modulated by various environmental stimuli, physio- and pathological conditions (Taupin, 2005). For example, environmental enrichment promotes the survival of newly generated neuronal cells in the DG. Voluntary running stimulates the generation of newly generated neuronal cells in the DG, but not the SVZ. Stress, neuroinflammation and aging decrease neurogenesis in the DG (Nithianantharajah and Hannan, 2006; Mora et al., 2007). In the diseased brain and after injuries to the CNS, like strokes and traumatic brain injuries (TBIs), neurogenesis is stimulated in the neurogenic areas, and new neuronal cells are generated at the sites of injuries, where they replace some of the degenerated nerve cells (Grote and Hannan, 2007). Cell tracking studies revealed that newly generated neuronal cells at sites of injuries originates from the SVZ. Newly generated neuronal cells migrate partially through the RMS to the degenerated areas. It is estimated that 0.2% of the degenerated nerve cells are replaced in the striatum after focal ischemia (Arvidsson et al., 2002). Hence, neurogenesis can be stimulated in the injured brain.

Limit and Pitfalls of BrdU Labeling

The modulation of neurogenesis and its quantification have been subject of debates, partly due to the use of BrdU, a thymidine analog, labeling as a method of assessment. As BrdU crosses the blood-brain barrier, it is generally administered intraperiteonally. It is suggested that activity, like exercise, but also the effects of various treatments and physio- and pathological conditions on cerebral flow, metabolism and permeability of the blood-brain barrier to reagents, and in particular to BrdU, may affect the availability of BrdU to the brain. The variation of BrdU quantification observed in these conditions would then reflect the change in BrdU uptake by the cells, rather than the modulation neurogenesis (Taupin, 2007).

With regard to the quantification of neurogenesis with BrdU, one study suggests that the standard concentration used to assess neurogenesis (50-100 mg/kg body weight in rodents, intraperitoneal injection) may not label all the dividing cells (Taupin, 2007), whereas another study reports that it does (Burns and Kuan, 2005). Further systematic studies on BrdU labeling in the CNS are thus needed to further define the conditions in which BrdU can be used for studying neurogenesis. The use of BrdU to study neurogenesis carries other limitations, like labeling of DNA repair, abortive cell cycle reentry and gene duplication, without cell proliferation (Taupin, 2007). Other strategies are therefore necessary to make educated conclusions with regard to adult neurogenesis when using BrdU labeling, like the study of markers of the cell cycle and use of retroviruses.

Mechanisms Underlying Adult Neurogenesis

Most the mechanisms underlying adult neurogenesis and NSC growth and fate determination are yet to be uncovered. It has been reported that cell death stimulates the proliferation of neural progenitor cells in the adult hippocampus (Gould and Tanapat, 1997). Other studies reveal that the mitotic rate is regulated by the number of available progenitor cells, rather than by cell death (Ekdahl et al., 2001; Jin et al., 2004). On the molecular level, epidermal growth factor and basic fibroblast growth factor were the first mitogens to be identified for neural progenitor and stem cells in vitro, and to stimulate neurogenesis in vivo (Reynolds and Weiss, 1992; Gage et al., 1995; Craig et al., 1996; Kuhn et al., 1997). However, other factors present in conditioned medium, like the glycosylated form of the protease inhibitor cystatin C (CCg), are also required for the proliferation of self-renewing, multipotent NSCs from single cells in vitro (Taupin et al., 2000), and remain to be characterized, as well as the pathways of these mitogens and cofactors.

Broader Potential of Adult Stem Cells

Adult stem cells are multipotents; they generate lineage specific cell types restricted to the tissues from which they are derived. Several studies have reported that adult-derived stem cells, and particularly adult-derived neural progenitor and stem cells, may have a broader potential; i.e., they generate cell types of lineages other than their tissues of origin (Bjornson et al., 1999; Brazelton et al., 2000; Mezey et al., 2000). However though some studies presented convincing results, phenomenon like contamination, transformation, transdifferentiation and cell fusion have been reported as possible explanation for the phenotypes observed in some studies (Anderson et al., 2001; Mezey, 2004).

Function(s) of Newborn Neuronal Cells

The function(s) of adult neurogenesis has been the source of intense research and debates. Evidences suggest that newly generated neuronal cells participate to process like learning and memory, and depression (Gould et al., 1999; Shors et al., 2001; Jacobs et al., 2000; Santarelli et al., 2003). The involvement of adult neurogenesis in learning and memory has been challenged by other studies. Increased hippocampal neurogenesis has been observed without improvement of learning and memory performances, in the Morris water maze test, in mice selectively bred for high levels of wheel running (Rhodes et al., 2003). Therefore the function of newly generated neuronal cells in the adult brain remains to be determined.

Finally, the evidence that neurogenesis occurs in the adult brain, and that NSCs reside in the adult CNS provide new avenues for cellular therapy. Cell therapeutic intervention may involve the stimulation of endogenous or the transplantation of neural progenitor and stem cells of the adult CNS. However, adult NSCs have yet to be brought to therapy.

Table 1. Adult neurogenesis and neural stem cells. Key publications

Year	Event (references)
1965	Seminal studies on adult neurogenesis (Altman and Das, 1965)
1982	Monoclonal antibody against BrdU (Gratzner, 1982)
1988	BrdU a marker to study neurogenesis (Miller and Nowakowski 1988)
1990	Identification of nestin as a marker for neural progenitor and stem cells (Lendahl et al., 1990)
1992	Post-mitotic cell death of newly generated neuronal cells in adult SVZ (Morshead and van der Kooy, 1992)
1992	Isolation and characterization of neural progenitor and stem cells from the adult rodent SVZ (Reynolds and Weiss, 1992)
1995	Isolation and characterization of neural progenitor and stem cells from the adult rodent hippocampus, adult hippocampal-derived neural progenitor and stem cells grafted in adult brain (Gage et al., 1995)
1997	Environmental enrichment promotes adult neurogenesis (Kempermann et al.,1997)
1998	Characterization of adult neurogenesis in the adult human hippocampus (Eriksson et al., 1998)
1999	Broader potential of adult-derived neural progenitor and stem cells (Bjornson et al., 1999)
1999	Glial origin for newly generated neural stem cells in the SVZ (Doetsch et al., 1999)
2001	Post-mitotic cell death of newly generated neuronal cells in adult hippocampus (Cameron and McKay, 2001)
2001	Isolation and characterization of neural progenitor and stem cells from human post-mortem tissues and biopsies (Palmer et al., 2001)
2005	Newly generated neuronal cells receive GABAergic excitatory input (Tozuka et al., 2005)
2007	Characterization of adult neurogenesis in the adult human SVZ (Curtis et al., 2007)

Though it is now accepted that neurogenesis occurs in the adult brain and NSCs reside in the adult CNS, much questions and controversies remain to be answered: what is the origin of newly generated neuronal cells in the adult brain, what are their molecular markers, what are the factors and mechanisms controlling NSC growth and fate specification, what is the potential of adult-derived stem cells, what are the functions of newly generated neuronal cells in the adult brain, and how can we use adult NSCs therapeutically?

The Future of Adult Neurogenesis

Newly generated neuronal cells represent a small fraction of nerve cells in the adult brain. But data presented above suggest that their relevance to CNS physio- and pathology, and cellular therapy as significant, but yet to be uncovered. One of the key underlyings of the importance of newly generated neuronal cells is their relative contributions compare to the preexisting network to CNS functioning. One can postulate that such contribution will depend on the specific properties of adult generated neuronal cells.

On the Functioning of Newly Generated Neuronal Cells in the Adult Brain

Adult newly generated neuronal cells belong to three groups based on their destinies. The first group consists of the newly generated neuronal cells that will undergo post-mitotic death (Morshead and van der Kooy, 1992; Cameron and McKay, 2001). The second group represents a population of newly generated cells that neither undergo apoptosis, nor differentiate to a defined fate. This latter group of cells likely contributes to renewing the stem cell niche. Niches are specialized microenvironments that regulate stem cells activity (Moore and Lemischka, 2006; Scadden, 2006). In the adult brain, neurogenic niches are maintained in restricted regions and have been identified and characterized (Alvarez-Buylla and Lim, 2004). These niches, an angiogenic and an astroglial niches, control NSCs self-renewal and differentiation (Palmer et al., 2000; Song et al., 2002). It is hypothesized that neurogenic niches underlie the properties and functions of NSCs in the adult CNS (Alvarez-Buylla and Lim, 2004; Taupin, 2006; Lim et al., 2007). The third group consists of the newly generated neuronal cells that will survive to maturity and integrate the network (Altman and Das, 1965; Eriksson et al., 1998; Kempermann et al., 2003; Dayer et al., 2003).

Several lines of evidence suggest that newly generated neuronal cells have different properties and physiological functions, than mature nerve cells, that may underlie their specific functions. Young granule cells in the adult DG appear to exhibit robust long-term potentiation that, in contrast to mature granule cells, cannot be inhibited by GABA (Wang et al., 2000). More recently, newly generated neuronal cells in the adult hippocampus were characterized as receiving GABAergic excitatory input (Ge et al., 2005, 2007; Tozuka et al., 2005), a function of GABA previously reported during development (Ben-Ari, 2002). Once cells have matured and integrated the pre-existing network, they may then functionally replace nerve cells born during development. Among the questions that arise from such theory are: what are the physiological functions of the newly generated neural cells during the time they are distinct from their mature counterpart, what is the function of such cellular renewal, and why would it occur only and specifically in discrete areas of the adult brain?

On the Functionality of Newly Generated Neuronal Cells in the Adult Brain

The increase of neurogenesis in diseases, disorders, and after injuries might then serve a neuroadaptative process (figure 1). Patients with neurological diseases, like Alzheimer's disease, epilepsy and Parkinson's disease (PD), but also recovering from strokes and injuries, are at greater risk of depression (Perna et al., 2003; Gilliam et al., 2004; Sawabini and Watts, 2004) and present memory impairments (Kotloski et al., 2002; Wang et al., 2004). Since learning and memory, and depression are associated with hippocampal neurogenesis (Gould et al., 1999; Jacobs et al., 2000; Shors et al., 2001; Santarelli et al., 2003), the depressive episode and learning impairments in patients suffering from neurological diseases, or disorders may contribute to the regulation of neurogenesis, in an additive, or cooperative manner with the disorder. Therefore, modulation of neurogenesis in the hippocampus might be an attempt by the CNS to compensate for other neuronal functions associated with the disease, like depression, and learning and memory impairments.

The increase in neurogenesis would also be a factor contributing to the plasticity of the CNS, and particularly related to the recovery in the CNS after injury. After cerebral strokes and TBIs, there is a striking amount of neurological recovery in the following months and years, despite often-permanent structural damage (Sbordone et al., 1995; Anderson et al., 2000). Though the mechanisms underlying such recovery are not fully understood, properties of plasticity of the CNS, like the reorganization of the pre-existing network and axonal sprouting have been implicated in the recovery (Ramic et al., 2006; Kolb and Gibb, 2007). Particularly, reorganization of the contra-lateral hemisphere has been involved in plasticity after brain injury (Cramer and Basting, 2000). Neurogenesis is increased bilaterally in the DG and the SVZ after cerebral strokes and TBIs. The bilateral increase in neurogenesis would contribute to the plasticity related recovery in the CNS, and particularly after injury.

The generation of newly generated neuronal cells at the sites of injury could represent a regenerative attempt by the CNS. In the diseased brain and after injuries to the CNS, new neuronal cells are generated at the sites of degeneration, where they replace some of the lost nerves cells (Arvidsson et al., 2002). Hence, there is no functional recovery. The generation of new neuronal cells at the sites of injury could represent an attempt by the CNS to regenerate following injury. Several hypotheses can explain the lack of recovery of the CNS after injury. The number of new neurons generated may be too low to compensate for the neuronal loss -0.2% of the degenerated nerve cells in the striatum after focal ischemia- (Arvidsson et al., 2002). The neuronal cells that are produced are non-functional because they do not develop into fully mature neurons, because they do not develop into the right type of neurons, or because they are incapable of integrating into the surviving brain circuitry. Gliogenesis has also been reported to occur at the sites of injuries (Fawcett and Asher, 1999). Therefore, neurogenesis and gliogenesis at the site of injuries may participate to a healing process.

The total number of neurons in the adult brain does not dramatically increase, and cell death is an established process in that adult brain (Gould et al., 2001). Newly generated neuronal cells may contribute to homeostasis of the tissue. Neurogenic niches have been described in the adult brain, and may hold the molecular and cellular cues to such phenomenon (Alvarez-Buylla and Lim, 2004). On the physiopathological level, an explanation is yet to be brought. It is worth mentioning that it has been suggested that since environmental enrichment promotes adult neurogenesis, and standard laboratory living condition do not represent physiological environment, neurogenesis may occur more broadly, at low level, that would remain undetected (Taupin, 2007), though such eventuality remains to be proven in mammals. Indeed, it has been proposed that self-repair mechanisms may operate in the adult rodent SN (Zhao et al., 2003), the area of the CNS affected in PD. If such turn-over of dopaminergic neuronal cells was confirmed, progression of the disease would then be determined not only by the rate of degeneration of SN neurons, but also by the efficacy in the formation of new dopamine neurons. Thus, disturbances of the equilibrium of cellular homeostasis could result in neurodegenerative diseases. So, in PD, neurogenesis might not only be a process for functional recovery, but it may also play a key role in the pathology of the disease. However, these data remain the source of controversies (Lie et al., 2002; Frielingsdorf et al., 2004), and such hypothesis remains to be demonstrated.

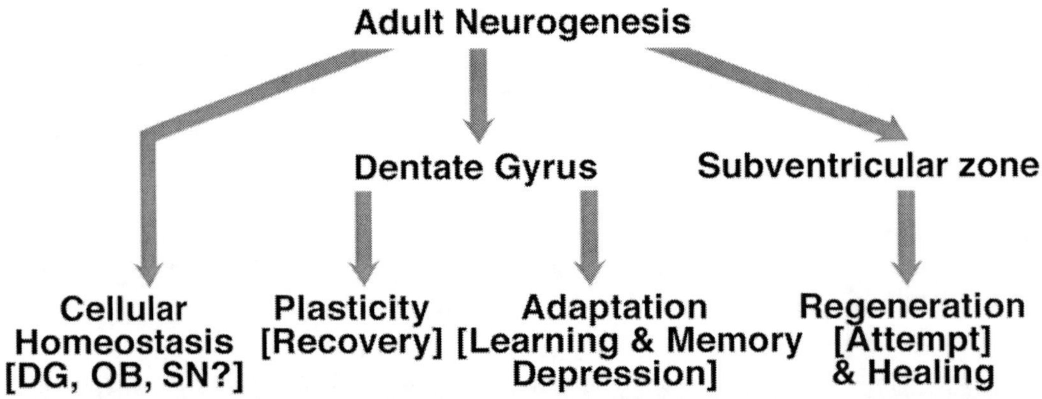

Figure 1. Functionality of adult neurogenesis. Adult neurogenesis occurs throughout adulthood. Hence the physiological function(s) of adult neurogenesis remains to be elucidated. Adult neurogenesis may be involved in the physiopathology of CNS functioning.

- Patients with neurological diseases, like Alzheimer's disease, epilepsy and Parkinson's disease, but also recovering from stroke and injury, are at greater risk of depression and present memory impairment. Since learning and memory, depression are associated with hippocampal neurogenesis, the increase of neurogenesis in diseases, disorders, and after injuries might then serve a neuroadaptative process.
- After cerebral strokes and traumatic brain injuries, there is a striking amount of neurological recovery in the following months and years, despite often-permanent structural damage. The increase in neurogenesis would also be a factor contributing to the plasticity of the CNS, and particularly related to the recovery in the CNS after injury.
- In the diseased brain and after injuries to the CNS, new neuronal cells are generated at the sites of degeneration, where they replace some of the lost nerves cells. The

generation of newly generated neuronal cells at the sites of injury could represent a regenerative attempt by the CNS, and its participation to the healing process.

- The total number of neurons in the adult brain does not dramatically increase, and cell death is an established process in that adult brain. Newly generated neuronal cells may contribute to cellular homeostasis. The desequilibrium in cellular homeostasis may result in neurodegenerative diseases.

The relative contribution of adult neurogenesis to these processes remains to be elucidated. Specific properties of newly generated neuronal cells yet to be determined would underlie the role of newly generated neuronal cells in CNS functioning.

Though at this time these hypotheses remain mostly speculative, the future of adult neurogenesis and NSC research lies in our understanding of the specific role and relative contribution of newly generated neuronal cells to the physio- and pathology of the CNS

The Promise of Adult Neural Stem Cells

The promise of adult NSCs lies also in our ability to bring adult NSC research to therapy. Because of their potential to generate the main phenotype of the CNS, NSCs hold the promise to cure a broad range of CNS diseases and injuries. The confirmation that neurogenesis occurs in the adult brain and NSCs reside in the adult CNS, opens new avenues for cellular therapy. Cell therapeutic intervention may involve the stimulation or grafting of neural progenitor and stem cells (Okano et al., 2007; Yamashima et al., 2007). The generation of new neuronal cells at the sites of injury further highlights the potential of the CNS to repair itself. The SVZ origin of the newly generated neuronal cells suggests that the stimulation of neurogenesis in the SVZ would provide a strategy to promote functional recovery after injury (Curtis et al., 2007b). Alternatively, the potential to isolate neural progenitor and stem cells from non-degenerated brain areas from the patient himself would provide an autologous source of transplantable neural progenitor and stem cells, thereby obviating the need to find a matching donor for the tissues and the use of drugs that suppress the immune system; thereby increasing the chance of successful graft and recovery. However such strategy would involve invasive surgery and the destruction of healthy brain tissue, a limiting factor for its clinical application. Neural progenitor and stem cells have also been isolated from human *post-mortem* tissues, providing an alternative source of tissues for cellular therapy (Palmer et al., 2001).

Conclusion

The promise of the future of adult neurogenesis and NSC research lies in our understanding of the function and relative contribution of newly generated neuronal cells in the adult brain, and our ability to bring adult NSC to therapy. The molecular, cellular, and physiological characterization of adult NSCs is a prerequisite to such endeavor. Significant advances have already been made in just the past decades. Because of the potential of adult neurogenesis and NSCs to redefine brain functioning, physio- and pathology, and its potential to cure a broad range of CNS diseases and injuries, the future of this field of research is tantalizing.

Acknowledgments

Reproduced, with permission from Dove Medical Press Ltd: Taupin P. Adult neural stem cells: the promise of the future. Neuropsychiatric Disease and Treatment (2007) 3(6):753-60.

References

Altman J. (1969) Autoradiographic and histological studies of postnatal neurogenesis. IV.Cell proliferation and migration in the anterior forebrain, with special reference to persisting neurogenesis in the olfactory bulb. *J. Comp. Neurol.* 137, 433-58.

Altman J, Das GD. (1965) Autoradiographic and histological evidence of postnatal hippocampal neurogenesis in rats. *J. Comp. Neurol.* 24, 319-6.

Alvarez-Buylla A, Lim DA. (2004) For the long run, maintaining germinal niches in the adult brain. *Neuron* 41, 683-6.

Anderson DJ, Gage FH, Weissman IL. (2001) Can stem cells cross lineage boundaries? *Nat. Med.* 7, 393-5.

Anderson VA, Catroppa C, Rosenfeld J, Haritou F, Morse SA. (2000) Recovery of intellectual ability following traumatic brain injury in childhood: impact of injury severity and age at injury. *Pediatr Neurosurg*. 32, 282-90.

Arvidsson A, Collin T, Kirik D, Kokaia Z, Lindvall O. (2002) Neuronal replacement from endogenous precursors in the adult brain after stroke. *Nat. Med.* 8, 963-70.

Ben-Ari Y. (2002) Excitatory actions of gaba during development: the nature of the nurture. *Nat. Rev. Neurosci.* 3, 728-39.

Brazelton TR, Rossi FM, Keshet GI, Blau HM. (2000) From marrow to brain: expression of neuronal phenotypes in adult mice. *Science* 290, 1775-9.

Bjornson CR, Rietze RL, Reynolds BA, Magli MC, Vescovi AL. (1999) Turning braininto blood: a hematopoietic fate adopted by adult neural stem cells in vivo. *Science* 283, 534-7.

Burns KA, Kuan CY. (2005) Low doses of bromo- and iododeoxyuridine produce near-saturation labeling of adult proliferative populations in the dentate gyrus. *Eur. J. Neurosci*. 21, 803-7.

Cameron HA, McKay RD. (2001) Adult neurogenesis produces a large pool of new granule cells in the dentate gyrus. *J. Comp. Neurol.* 435, 406-17.

Carlen M, Cassidy RM, Brismar H, Smith GA, Enquist LW, Frisen J. (2002) Functional integration of adult-born neurons. *Curr. Biol.* 12, 606-8.

Cramer SC, Bastings EP. (2000) Mapping clinically relevant plasticity after stroke. *Neuropharmacology* 39, 842-51.

Craig CG, Tropepe V, Morshead CM, Reynolds BA, Weiss S, van der Kooy D. (1996) In vivo growth factor expansion of endogenous subependymal neural precursor cell populations in the adult mouse brain. *J. Neurosci.* 16, 2649-58.

Curtis MA, Kam M, Nannmark U, Anderson MF, Axell MZ, Wikkelso C, Holtås S, van Roon-Mom WM, Björk-Eriksson T, Nordborg C, Frisén J, Dragunow M, Faull RL, Eriksson PS. (2007a) Human neuroblasts migrate to theolfactory bulb via a lateral ventricular extension. *Science* 315, 1243-9.

Curtis MA, Eriksson PS, Faull RL. (2007b) Progenitor cells and adult neurogenesis in neurodegenerative diseases and injuries of the basal ganglia. *Clin. Exp. Pharmacol. Physiol.* 34, 528-32.

Dayer AG, Ford AA, Cleaver KM, Yassaee M, Cameron HA. (2003) Short-term and long-term survival of new neurons in the rat dentate gyrus. *J. Comp. Neurol.* 460, 563-72.

Doetsch F, Caille I, Lim DA, Garcia-Verdugo JM, Alvarez-Buylla A. (1999) Subventricular zone astrocytes are neural stem cells in the adult mammalian brain. *Cell* 97, 703-16.

Ekdahl CT, Mohapel P, Elmer E, Lindvall O. (2001) Caspase inhibitors increase short-term survival of progenitor-cell progeny in the adult rat dentate gyrus following status epilepticus. *Eur. J. Neurosci.* 14, 937-45.

Eriksson PS, Perfilieva E, Björk-Eriksson T, Alborn AM, Nordborg C, Peterson DA, Gage FH. (1998) Neurogenesis in the adult human hippocampus. *Nat. Med.* 4, 1313-7.

Fawcett JW, Asher RA. (1999) The glial scar and central nervous system repair. *Brain Res. Bull.* 49, 377-91.

Filippov V, Kronenberg G, Pivneva T, Reuter K, Steiner B, Wang LP, Yamaguchi M, Kettenmann H, Kempermann G. (2003) Subpopulation of nestin-expressing progenitor cells in the adult murine hippocampus shows electrophysiological and morphological characteristics of astrocytes. *Mol. Cell Neurosci.* 23, 373-82.

Frielingsdorf H, Schwarz K, Brundin P, Mohapel P. (2004) No evidence for new dopaminergic neurons in the adult mammalian substantia nigra. *Proc. Natl. Acad. Sci. USA* 101, 10177-82.

Gage FH, Coates PW, Palmer TD, Kuhn HG, Fisher LJ, Suhonen JO, Peterson DA, Suhr ST, Ray J. (1995) Survival and differentiation of adult neuronal progenitor cells transplanted to the adult brain. *Proc. Natl. Acad. Sci. USA* 92, 11879-83.

Garcia AD, Doan NB, Imura T, Bush TG, Sofroniew MV. (2004) GFAP-expressing progenitors are the principal source of constitutive neurogenesis in adult mouse forebrain. *Nat. Neurosci.* 7, 1233-41.

Ge S, Goh EL, Sailor KA, Kitabatake Y, Ming GL, Song H. (2005) GABA regulates synaptic integration of newly generated neurons in the adult brain. *Nature* 439, 589-93.

Ge S, Pradhan DA, Ming GL, Song H. (2007) GABA sets the tempo for activity-dependent adult neurogenesis. *Trends Neurosci.* 30, 1-8.

Gilliam FG, Santos J, Vahle V, Carter J, Brown K, Hecimovic H. (2004) Depression in epilepsy: ignoring clinical expression of neuronal network dysfunction? *Epilepsia* 45, 28-33.

Gould E. (2007) How widespread is adult neurogenesis in mammals? *Nat. Rev. Neurosci.* 8, 481-8.

Gould E, Tanapat P. (1997) Lesion-induced proliferation of neuronal progenitors in the dentate gyrus of the adult rat. *Neurosci.* 80, 427-36.

Gould E, Beylin A, Tanapat P, Reeves A, Shors TJ. (1999) Learning enhances adult neurogenesis in the hippocampal formation. *Nat. Neurosci*. 2, 260-5.

Gould E, Reeves AJ, Graziano MS, Gross CG. (1999) Neurogenesis in the neocortex of adult primates. *Science* 286, 548-52.

Gould E, Vail N, Wagers M, Gross CG. (2001) Adult-generated hippocampal and neocortical neurons in macaques have a transient existence. *Proc. Natl. Acad. Sci. USA* 98, 10910-7.

Gratzner HG. (1982) Monoclonal antibody to 5-bromo- and 5-iododeoxyuridine: A new reagent for detection of DNA replication. *Science* 218, 474-5.

Gross CG. (2000) Neurogenesis in the adult brain: death of a dogma. *Nat. Rev. Neurosci.* 1, 67-73.

Grote HE, Hannan AJ. (2007) Regulators of adult neurogenesis in the healthy and diseased brain. *Clin. Exp. Pharmacol. Physiol*. 34, 533-45.

Jacobs BL, Praag H, Gage FH. (2000) Adult brain neurogenesis and psychiatry: a novel theory of depression. *Mol. Psychiatry* 5, 262-9.

Jin K, Galvan V, Xie L, Mao XO, Gorostiza OF, Bredesen DE, Greenberg DA. (2004) Enhanced neurogenesis in Alzheimer's disease transgenic (PDGF-APPSw,Ind) mice. *Proc. Natl. Acad. Sci. USA* 101, 13363-7.

Kaneko Y, Sakakibara S, Imai T, Suzuki A, Nakamura Y, Sawamoto K, Ogawa Y, Toyama Y, Miyata T, Okano H. (2000) Musashi1: an evolutionally conserved marker for CNS progenitor cells including neural stem cells. *Dev. Neurosci.* 22, 139-53.

Kato T, Yokouchi K, Fukushima N. (2001) Continual replacement of newly-generated olfactory neurons in adult rats. *Neurosci. Lett.* 307, 17-20.

Kempermann G, Kuhn HG, Gage FH. (1997) More hippocampal neurons in adult mice living in an enriched environment. *Nature* 386, 493-5.

Kempermann G, Gast D, Kronenberg G, Yamaguchi M, Gage FH. (2003) Early determination and long-term persistence of adult-generated new neurons in the hippocampus of mice. *Development* 130, 391-9.

Kolb B, Gibb R. (2007) Brain plasticity and recovery from early cortical injury. *Dev. Psychobiol.* 49, 107-18.

Komitova M, Eriksson PS. (2004) Sox-2 is expressed by neural progenitors and astroglia in the adult rat brain. *Neurosci. Lett*. 369, 24-7.

Kornack DR, Rakic P. (2001) Cell proliferation without neurogenesis in adult primate neocortex. *Science* 294, 2127-30.

Kotloski R, Lynch M, Lauersdorf S, Sutula T. (2002) Repeated brief seizures induce progressive hippocampal neuron loss and memory deficits. *Prog. Brain Res*. 135, 95-110.

Kuhn HG, Winkler J, Kempermann G, Thal LJ, Gage FH. (1997) Epidermal growth factor and fibroblast growth factor-2 have different effects on neural progenitors in the adult rat brain. *J. Neurosci.* 17, 5820-9.

Lendahl U, Zimmerman LB, McKay RD. (1990) CNS stem cells express a new class of intermediate filament protein. *Cell* 60, 585-95.

Lie DC, Dziewczapolski G, Willhoite AR, Kaspar BK, Shults CW, Gage FH. (2002) The adult substantia nigra contains progenitor cells with neurogenic potential. *J. Neurosci.* 22, 6639-49.

Lim DA, Huang YC, Alvarez-Buylla A. (2007) The adult neural stem cell niche: lessons for future neural cell replacement strategies. *Neurosurg. Clin. N. Am.* 18, 81-92, ix.

Mezey E, Chandross KJ, Harta G, Maki RA, McKercher SR. (2000) Turning blood into brain: cells bearing neuronal antigens generated in vivo from bone marrow. *Science* 290, 1779-82.

Mezey E. (2004) Commentary: on bone marrow stem cells and openmindedness. *Stem Cells Dev.* 13, 147-52.

Miller MW, Nowakowski RS. (1988) Use of bromodeoxyuridine-immunohistochemistry to examine the proliferation, migration and time of origin of cells in the central nervous system. *Brain Res.* 457, 44-52.

Moore KA, Lemischka IR. (2006) Stem cells and their niches. *Science* 311, 1880-5.

Mora F, Segovia G, Del Arco A. (2007) Aging, plasticity and environmental enrichment: Structural changes and neurotransmitter dynamics in several areas of the brain. *Brain Res. Rev.* 55, 78-88.

Morshead CM, van der Kooy D. (1992) Postmitotic death is the fate of constitutively proliferating cells in the subependymal layer of the adult mouse brain. *J. Neurosci.* 12, 249-56.

Nithianantharajah J, Hannan AJ. (2006) Enriched environments, experience-dependent plasticity and disorders of the nervous system. *Nat. Rev. Neurosci.* 7, 697-709.

Okano H, Sakaguchi M, Ohki K, Suzuki N, Sawamoto K. (2007) Regeneration of the central nervous system using endogenous repair mechanisms. *J. Neurochem.* 102, 1459-65.

Okuda T, Tagawa K, Qi ML, Hoshio M, Ueda H, Kawano H, Kanazawa I, Muramatsu M, Okazawa H. (2004) Oct-3/4 repression accelerates differentiation of neural progenitor cells in vitro and in vivo. *Brain Res. Mol. Brain Res.* 132, 18-30.

Palmer TD, Willhoite AR, Gage FH. (2000) Vascular niche for adult hippocampal neurogenesis. *J. Comp. Neurol.* 425, 479-94.

Palmer TD, Schwartz PH, Taupin P, Kaspar B, Stein SA, Gage FH. (2001) Cell culture. Progenitor cells from human brain after death. *Nature* 411, 42-3.

Perna RB, Rouselle A, Brennan P. (2003) Traumatic brain injury: depression, neurogenesis, and medication management. *J. Head Trauma Rehabil.* 18, 201-3.

Potten CS, Loeffler M. (1990) Stem cells: attributes, cycles, spirals, pitfalls and uncertainties. Lessons for and from the crypt. *Development* 110, 1001-20.

Ramic M, Emerick AJ, Bollnow MR, O'Brien TE, Tsai SY, Kartje GL. (2006) Axonal plasticity is associated with motor recovery following amphetamine treatment combined with rehabilitation after brain injury in the adult rat. *Brain Res.* 1111, 176-86.

Reynolds BA, Weiss S. (1992) Generation of neurons and astrocytes from isolated cells of the adult mammalian central nervous system. *Science* 255, 1707-10.

Rhodes JS, van Praag H, Jeffrey S, Girard I, Mitchell GS, Garland T Jr, Gage FH. (2003) Exercise increases hippocampal neurogenesis to high levels but does notimprove spatial learning in mice bred for increased voluntary wheel running. *Behav. Neurosci.* 117, 1006-16. Erratum in: (2003) *Behav. Neurosci.* 118, 305.

Rietze R, Poulin P, Weiss S. (2000) Mitotically active cells that generate neurons and astrocytes are present in multiple regions of the adult mouse hippocampus. *J. Comp. Neurol.* 424, 397-408.

Sakakibara S, Imai T, Hamaguchi K, Okabe M, Aruga J, Nakajima K, Yasutomi D, Nagata T, Kurihara Y, Uesugi S, Miyata T, Ogawa M, Mikoshiba K, Okano H. (1996) Mouse-Musashi-1, a neural RNA-binding protein highly enriched in the mammalian CNS stem cell. *Dev. Biol.* 176, 230-42.

Santarelli L, Saxe M, Gross C, Surget A, Battaglia F, Dulawa S, Weisstaub N, Lee J, Duman R, Arancio O, Belzung C, Hen R. (2003) Requirement of hippocampal neurogenesis for the behavioral effects of antidepressants. *Science* 301, 805-9.

Sawabini KA, Watts RL. (2004) Treatment of depression in Parkinson's disease. *Parkinsonism Relat. Disord.* 10, S37-41.

Sbordone RJ, Liter JC, Pettler-Jennings P. (1995) Recovery of function following severe traumatic brain injury: a retrospective 10-year follow-up. *Brain Inj.* 9, 285-99.

Scadden DT. (2006) The stem-cell niche as an entity of action. *Nature* 441, 1075-9.

Shors TJ, Miesegaes G, Beylin A, Zhao M, Rydel T, Gould E. (2001) Neurogenesis in the adult is involved in the formation of trace memories. *Nature* 410, 372-6. Erratum in: (2001) *Nature* 414, 938.

Song H, Stevens CF, Gage FH. (2002) Astroglia induce neurogenesis from adult neural stem cells. *Nature* 417, 39-44.

Taupin P, Ray J, Fischer WH, Suhr ST, Hakansson K, Grubb A, Gage FH. (2000) FGF-2-responsive neural stem cell proliferation requires CCg, a novel autocrine/paracrine cofactor. *Neuron* 28, 385-97.

Taupin P, Gage FH. (2002) Adult neurogenesis and neural stem cells of the central nervous system in mammals. *J. Neurosci. Res.* 69, 745-9.

Taupin P. (2005) Adult neurogenesis in the mammalian central nervous system: functionality and potential clinical interest. *Med. Sci. Monit.* 11, RA247-52.

Taupin P. (2006) Adult neural stem cells, neurogenic niches and cellular therapy. *Stem Cell Reviews* 2, 213-20.

Taupin P. (2007) BrdU Immunohistochemistry for Studying Adult Neurogenesis: paradigms, pitfalls, limitations, and validation. *Brain Res. Rev.* 53, 198-214.

Tozuka Y, Fukuda S, Namba T. (2005) GABAergic excitation promotes neuronal differentiation in adult hippocampal progenitor cells. *Neuron* 47, 803-15.

van Praag H, Schinder AF, Christie BR, Toni N, Palmer TD, Gage FH. (2002) Functional neurogenesis in the adult hippocampus. *Nature* 415, 1030-4.

Wang LP, Kempermann G, Kettenmann H. (2005) Subpopulation of precursor cells in the mouse dentate gyrus receives synaptic GABAergic input. *Mol. Cell Neurosci.* 29, 181-9.

Wang R, Dineley KT, Sweatt JD, Zheng H. (2004) Presenilin 1 familial Alzheimer's disease mutation leads to defective associative learning and impaired adult neurogenesis. *Neurosci.* 126, 305-12.

Wang S, Scott BW, Wojtowicz JM. (2000) Heterogenous properties of dentate granule neurons in the adult rat. *J. Neurobiol.* 42, 248-57.

Yamashima T, Tonchev AB, Yukie M. (2007) Adult hippocampal neurogenesis in rodents and primates: endogenous, enhanced, and engrafted. *Rev. Neurosci.* 18, 67-82.

Zappone MV, Galli R, Catena R, Meani N, De Biasi S, Mattei E, Tiveron C, Vescovi AL, Lovell-Badge R, Ottolenghi S, Nicolis SK. (2000) Sox2 regulatory sequences direct expression of a (beta)-geo transgene to telencephalic neural stem cells and precursors of the mouse embryo, revealing regionalization of gene expression in CNS stem cells. *Development* 127, 2367-82.

Zhao M, Momma S, Delfani K, Carlen M, Cassidy RM, Johansson CB, Brismar H, Shupliakov O, Frisen J, Janson AM. (2003) Evidence for neurogenesis in the adult mammalian substantia nigra. *Proc. Natl. Acad. Sci. USA* 100, 7925-30.

Chapter I

Neurogenesis in the Adult Central Nervous System

Abstract

Contrary to the long-held dogma, neurogenesis occurs throughout adulthood, and neural stem cells reside in the adult central nervous system (CNS) in mammals. The developmental process of the brain may thus never end, and the brain may be amenable to repair. Neurogenesis is modulated in a wide variety of physiological and pathological conditions, and is involved in processes such as learning and memory and depression. However, the relative contribution of newly generated neuronal cells to these processes, as well as to CNS plasticity, remains to be determined. Thus, not only neurogenesis contributes to reshaping the adult brain, it will ultimately lead us to redefine our knowledge and understanding of the nervous system.

Introduction

Neural stem cells (NSCs) are the self-renewing, multipotent participate to the formation of the nervous system. It was believed that the generation of neuronal cells in mammals was mostly limited to the pre-natal phase of development, and that the adult brain was devoid of stem cells, and thus of the ability to make new nerve cells and regenerate after injuries [1]. Seminal studies in the 1960s that were substantiated in the 1970s and 1980s, reported that neurogenesis occurs in discrete areas of the adult brain in rodents [2-6]. With the advent of new methods for labeling dividing cells, such as 5-bromo-2'-deoxyuridine (BrdU, figure 1) labeling, retroviral labeling and confocal microscopy, investigators have confirmed that neurogenesis occurs in discrete areas of the rodent brain throughout adulthood [7-12], and reassessed and presented evidences that adult neurogenesis also occurs in primates, humans and non-humans [13-15]. BrdU is a thymidine analogue used for birth-dating and monitoring cell proliferation [16]. BrdU is generally administered intraperitoneally; it inserts into the DNA of dividing cells, including in the central nervous system (CNS), as it crosses the blood-brain barrier [17]. It is hypothesized that newly generated neuronal cells originate from stem cells in the adult brain. A hypothesis further supported by the recent isolation and characterization of neural progenitor and stem cells from the adult brain [18, 19]. The confirmation that neurogenesis occurs in the adult brain and NSCs reside in the adult CNS have profound implications for our understanding of brain development and functioning, as well as for cellular therapy in the CNS.

Neurogenesis in the Adult Brain

Neurogenesis occurs in two forebrain regions of the adult brain: the subventricular zone (SVZ) and the dentate gyrus (DG) in various species [20]. Of these two regions, the SVZ harbours the largest pool of dividing neural progenitor cells in the adult rodent brain [21, 22]. Newly generated neuronal cells in the SVZ migrate to the olfactory bulb (OB) through the rostro-migratory stream (RMS), where they differentiate into granule and periglomerular neurons of the OB, in rodents [4, 10, 23] and in macaque monkeys [15, 24] (figure 1a). Macaque monkeys are nonhuman Old World anthropoids, phylogenetically close to humans. Like humans they have a hippocampal formation, a relatively small OB – compared with rodents –, a life-history pattern and are largely diurnal [25, 26]. Progenitor cells in the SVZ aggregate to form an extensive network of neuroblast chains that migrate tangentially through the SVZ and coalesce anteriorly to form the RMS [27, 28]. The network of chains of migrating neuroblasts in the SVZ and RMS can be revealed by immunostaining for polysialylated neural cell adhesion molecule (PSA–NCAM) [15, 29]. Neuroblasts undergoing 'chain migration' along the SVZ/RMS migrate along one another via homotypic interactions, without radial glial or axonal guides [30]. The directional migration of neuroblasts along the SVZ requires the beating of the ependymal cilia [31]. In rodents (mice), a SVZ progenitor cell requires at least 15 days to be generated, migrate 3–5 mm and differentiate into new olfactory interneurons. These cells migrate at an average rate of 30 mmh^{-1} [10]. In macaque monkeys, a SVZ progenitor cell requires at least 75 to 97 days to be generated, migrate 20 mm and differentiate into new olfactory interneurons, a process slower than in rodents [15]. In the DG, newly generated neuronal cells in the subgranular zone (SGZ) migrate to the granule layer, where they differentiate into neuronal cells and extend axonal projections to the CA3 area of the Ammon's horn in rodents [3, 5-7, 11, 32, 33], in New-World monkeys (marmoset) [13] and in Old-World monkeys [34, 35] (figure 1c–e). In rodents (rats), immature granule cells extend axons into CA3 as rapidly as 4-10 days after mitosis [36], whereas the maturation of newborn cells, from the proliferation of newly generated cells in the SGZ to the migration and differentiation in neuronal cells of the granule cell layer, takes approximately four weeks [7]. As many as 9,000 new neuronal cells are generated per day in the rodent DG, contributing to about 3.3% per month or about 0.1% per day of the granule cell population [37, 38], whereas in adult macaque monkey, it is estimated that at least 0.004% of the neuronal population in the granule cell layer are new neurons generated per day [35].The relative rate of neurogenesis is estimated to be approximately 10 times less in adult macaque monkeys than that reported in the adult rodent DG [35]. However, it is difficult to make a quantitative comparison between the species, as nothing is known of the relative bioavailability of BrdU or its ability to cross the blood–brain barrier and be available for uptake by dividing cells in the two species. Neurogenesis may also occur in other areas of the adult brain – albeit at lower level –, such as the CA1 area [39], the neocortex [40, 41], the striatum [42], the amygdala [43], the substantia nigra [44], the third ventricle [45], the subcortical white matter [46], the caudate nucleus [47] in some species [19-25]. However, some of these data have been the sources of debates and controversies and remain to be further confirmed [48-50].

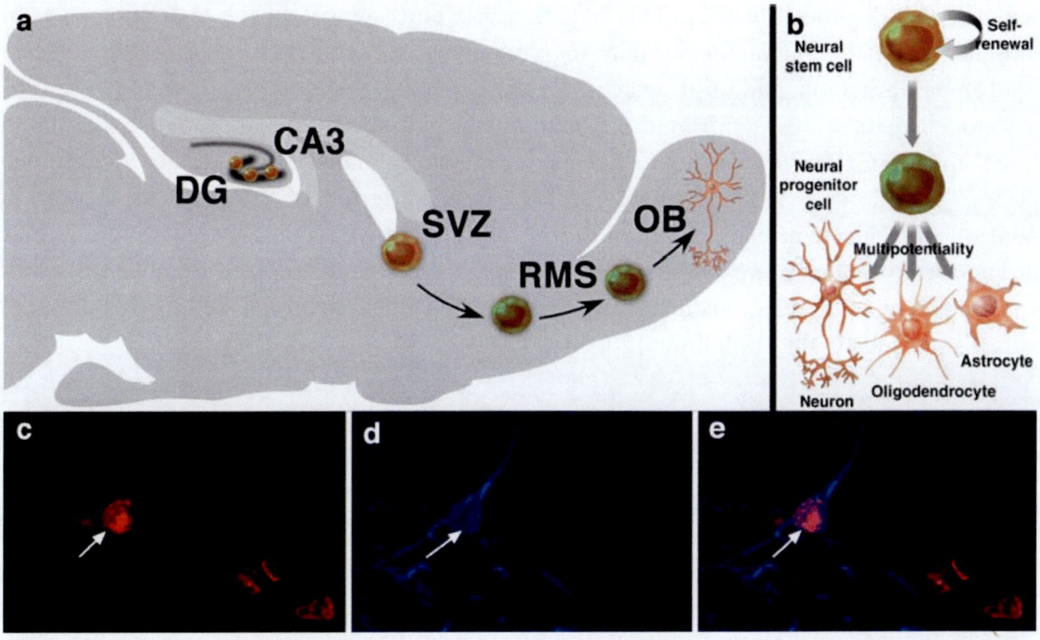

Figure 1. Neurogenesis and neural stem cells in the adult brain. Neurogenesis occurs mainly in two areas of the adult brain, the DG and the SVZ. In the DG, new neuronal cells are generated in the subgranular zone. In the SVZ, newly generated neuronal cells migrate to the OB, through the RMS, where they differentiate into interneurons of the OB (a). NSCs are the self-renewing, multipotent cells that generate the neuronal and glial cells of the nervous system. Neural progenitor cells are multipotent cells with limited proliferative capacity (b). 5-Bromo-2'-deoxyuridine-labeling is the standard for studying neurogenesis and its regulation. Co-labeling of a BrdU-positive cell (c, red, arrow) with class III ß-tubulin isotype (d, blue) in the DG. Tuj-1 is a marker of immature neuronal cells. The merge picture shows a BrdU-positive cell also positive for Tuj-1 (arrow, e) in the DG, representative of a newly generated neuronal cell.

In the human, Eriksson et al. [14] performed immunofluorescence labeling for BrdU and neuronal markers on brain tissue samples obtained post-mortem from cancer patients who had been treated with BrdU to assess the proliferative activity of the tumour cells, and revealed that neurogenesis occurs in the DG in the adults [14]. In the SVZ, Sanai et al. (2004) performed immunofluorescence labeling for the cell cycle marker Ki67 on brain tissue specimens from neurosurgical resections and autopsied brains, and reported that the adult human brain contains a ribbon of astrocytes lining the lateral ventricles, a particular organization not observed in other vertebrates, including nonhuman primates. Some cells in the astrocytes ribbon divide, as revealed by the co-expression of glial fibrillary acidic protein (GFAP) and the cell-division marker Ki-67. However, Sanai et al. found no evidence of chains of migrating neuroblasts in the SVZ or in the pathway to the OB. These results suggest that neurons in the adult human OB are not replaced [51]. Alternatively, precursors may migrate as individual cells, or the pathway may be different in the human SVZ and yet to be found. But it highlights the particularity of neurogenesis in the adult human brain. The rate of neurogenesis in the human DG was also reported to be low [14]. The reasons for the apparent reduction of neurogenesis in adult primates are unclear. The decline of adult neurogenesis during vertebrate evolution could be an adaptative strategy to maintain stable neuronal populations throughout life [52]. This hypothesis is consistent with the restriction of adult neurogenesis in the mammalian brain to phylogenetically older structures, the OB and hippocampal formation, and its absence in the more recently evolved neocortex [48, 52, 53].

Altogether these data confirmed that neurogenesis occurs in the adult mammalian brain, though the particularities of neurogenesis in human remain to be further determined. Newborn granule cells in the DG survive for extended period of time, at least 2 years in humans [14]. Thus, neuronal cells born during adulthood that becomeintegrated into circuits and survive to maturity are very stable, and may permanently replace granule cells born during development. It is postulated that newly generated neuronal cells originate from residual stem cells in the adult brain [20, 54].

Neural Progenitor and Stem Cells from the Adult Brain

In the adult brain, it is hypothesized that NSCs reside in the adult SVZ and hippocampus. In 1992, Reynolds and Weiss were the first to isolate and characterize a population of undifferentiated cells or neural progenitor cells, positive for the neural progenitor and stem cell marker nestin [55], from the adult mice striatal area, including the SVZ [18]. In 1995, Gage et al. isolated a similar population of cells, from the adult rat hippocampus [19]. One of the limitations with the established cultures derived from the adult brain is that they are heterogeneous; they contain neural progenitor and stem cells. Neural progenitor cells are defined as multipotent cells, with limited proliferative capacity [20, 56, 57] (figure 1b). Further, there is no specific marker of NSCs *in vitro* and *in vivo*; the intermediate filament nestin, a marker for neural progenitor and stem cells, is also expressed *in vivo* in reactive astrocytes and gliomas [58, 59]. Other markers of neural progenitor and stem cells, such as the transcription factors sox-2, oct-3/4, and the RNA binding protein Musashi 1, also label population of glial cells [60-64]. Despite intense effort, NSCs are yet to be identified [65-67]. Hence, the characterization *in vitro* of cells with NSC properties requires demonstrate self-renewal over an extended period of time (more than five passages), multipotentiality, coincident with the generation of a large number of progeny, several orders of magnitude more numerous than the starting population [68, 69]. In 1996, Gritti et al. isolated self-renewal, multipotent NSCs from cultures derived from the SVZ, and in 1997, Palmer et al. isolated and characterized self renewal, multipotent NSCs derived from the adult hippocampus [70, 71]. Recent data have challenged the existence of NSCs in the adult hippocampus, reporting the isolation of neural progenitor, but not stem cells from adult hippocampus [72, 73]. However, differences in protocols, species, handling technique may explain discrepancies between the studies, and such claim that self-renewal, multipotent NSCs cannot be isolated from the adult hippocampus would require comparing the two models in a strict experimental approach. Altogether, these data suggest that NSCs reside in the adult SVZ and hippocampus. Neural progenitor and stem cells have also been isolated and characterized *in vitro* form other areas of the adult brain and the spinal cord, suggesting that neural progenitor and stem cells may also reside in other areas of the CNS, where they would be quiescent [20]. The isolation and characterization of neural progenitor and stem cells from the adult brain provide a source of tissue for cellular therapy.

Neural Progenitor and Stem Cells in the Adult Brain

Despite the lack of molecular markers for NSCs, researchers have aimed at identifying the origin of newly generated neuronal cells in the adult brain. Based on ultrastructure, cell cycle analysis, [^{3}H]-thymidine autoradiography and immunocytochemical studies, particularly BrdU-labeling, two conflicting theories with regard to the origin of newly generated neuronal cells in the adult brain have been proposed. One theory contends that newly generated neuronal cells originate from astrocyte-like cells, expressing the glial marker, GFAP, and nestin, in the SVZ [74]. This finding was surprising, because astrocytes are considered as differentiated cells belonging to the glial lineage. The second theory contends that newly generated neuronal cells originate from a population of ependymal cells in the SVZ that express the intermediate filament protein nestin [75]. The ependyma, as the subependyma, originate from the embryonic forebrain germinal zones. However, this conclusion is not supported by other studies [76, 77], and remains to be further confirmed [20]. In the hippocampus, Seri et al. [78] and Fillippov et al. [79], using similar approaches, as well as retroviral labeling, studies in transgenic mice expressing the green fluorescent protein (GFP) under the nestin promoter and electrophysiology, showed that SGZ astrocytes are the primary precursors in the formation of new neurons in the adult hippocampus [78, 79].

Three main cell types relevant to adult neurogenesis in the SVZ have been identified and characterized in the SVZ, in rodents [80, 81] and in macaques [15]. Type-A cells are elongated and smooth cells that do not express GFAP and the intermediate filament vimentine. Type-A cells are immunopositive for nestin, PSANCAM-, and ß-tubulin (Tuj1). Type-A cells course tangentially to the walls of the lateral ventricle throughout the SVZ [28]. They incorporate [^{3}H]-thymidine, and form chains of migrating cells [27]. Type-A cells correspond to the migrating neural precursors or neuroblasts. Type-B cells have irregular shapes. They express nestin, GFAP and vimentine, but are immunonegative for PSA-NCAM and Tuj1. Type-B cells are detected in the SVZ, where they form a tubular trabecula that ensheathes the chains of type-A cells, isolating them from surrounding parenchyma [27]. Type-B cells are astrocytes, and may correspond to the stem cells *in vivo*. Type-B cells would correspond to the population of relative quiescent cells previously proposed as NSCs in the SVZ [82]. Unlike in the RMS, two types of B cells are distinguished in the SVZ: type B1 and B2. Type B1 astrocytes separate the chains of type-A cells from the ependymal layer, whereas the basally located type B2 cells separate the chains of type-A cells from the surrounding striatal parenchyma. Type-B2 (but not B1) cells incorporated [^{3}H]-thymidine. Type-C cells are of large size, have a smooth contour, express nestin, and are immunonegative for GFAP, vimentine, PSA-NCAM and Tuj1. Type-C cells are present throughout the lateral wall of the SVZ, but not in the RMS, where they are often found in clusters, closely associated to the migrating neuroblasts or chains of type-A cells. Type-C cells are the most actively proliferating cells in the SVZ (50% of [^{3}H]-thymidine-labelled cells were reported as type-C cells). Type-C cells may generate both neurons and glia, and correspond to precursors of the type-A cells, the migrating neuroblasts. From these data, Doetsch et al. [80] proposed a model for neurogenesis in adult the SVZ: type B cells that are NSCs, give rise to neural precursors type C cells in the adult SVZ, that in turn generate type-A cells that migrate in chains through glial tubes formed by SVZ astrocytes (type-B cells) along the RMS, and will differentiate into interneurons in the OB [80].This model has been confirmed after studying the order of reappearance of different SVZ cell types and their proliferation, after antimitotic treatment with cytosine-ßD-arabinofuranoside (Ara-C). After Ara-C treatment, migrating neuroblasts (type-A cells), and type-C cells are eliminated, but some astrocytes, type-B cells and ependymal cells remain. The SVZ network then rapidly regenerates: first, astrocytes type-B cells divide, two days later, type-C cells reappear, followed at 4.5 days by migrating neuroblasts. By 10 days, the SVZ network is fully regenerated, as in normal mice. These data suggest that Ara-C treatment kills neuronal precursors cells, type-A and -C cells that are actively dividing cells, but spare the NSCs, type-B cells, that were able to regenerate [83].

In the hippocampus, Seri et al. [78] described a population of small electron-dense cells in the SGZ, type-D cells that are of small size, immunonegative for GFAP and are dividing. Type-D cells probably function as transient precursors in the formation of new neurons, as anti-mitotic treatment resulted in the elimination of D cells from the SGZ. It is proposed that SGZ astrocytes, putative NSCs, would give rise to type-D cells, which in turn would give rise to newly generated neuronal cells in the granule cell layer [78]. Type-D cells in the SGZ do not divide as frequently as type-C cells in the SVZ, suggesting that the amplification for neuronal production by transient precursors in the SGZ is probably limited [78, 84]. Filippov et al. [79] further characterized the nestin positive cells in the adult hippocampus of nestin-GFP transgenic mice [79]. Filippov et al. described two sub-populations of nestin-GFP-positive cells. Type-1 cells have long processes, with vascular end feet, express GFAP, and have electrophysiological features of astrocytes – cells with delayed-rectifying potassium currents. Type-1 cells correspond to differentiated astrocytes, presumably corresponding to the type-B cells as described by Seri et al. [78]. Type-2 cells are small in size, express GFAP, are immunonegative for S100-ß, a marker for astrocytes [85], and have electrophysiological features of the earliest steps of neuronal differentiation – cells with sodium currents. Type-2 cells lacking astrocytic features would correspond to the D cells described by Seri et al. [78]. These data further support the hypothesis that NSCs in the hippocampus correspond to a sub-population of astrocytes, and that nestin expressing cells are heterogeneous in the adult brain. These results also indicate that GFAP, an intermediate neurofilament, is also expressed by cells with early neuronal features in the adult DG. Therefore, a population of progenitor cells expresses GFAP, whether or not this alone qualifies them as glial cells or astrocytes.

Altogether, these data suggest that a population of astrocyte-like cells correspond to NSCs in the adult SVZ and hippocampus. Though the a glial origin for newly generated neuronal cells in the adult brain has received much support recently [86-88], the origin and identity of newly generated neuronal cells in the adult brain are still the subject to debates and controversies, and remain to be further evaluated [20]. Song et al. [89] recently reported that astrocytes in the adult brain promote neurogenesis [89]. The proliferation of astrocytes following Ara-C treatment could stimulate the proliferation of a yet unidentified NSC that is slowly dividing and/or is resistant to Ara-C treatment. It also remains to determine unambiguously the relationship between the precursors of newly generated neuronal cells *in vivo*, and the neural progenitor and stem cells derived from the adult brain and cultured *in vitro*. The identification of neural progenitor and stem cells in the adult brain suggests the adult brain has the potential for self-repair.

A Developmental Process

During development, gamma aminobutyric acid (GABA) has been reported to act as an excitatory neurotransmitter on embryonic neural progenitor cells, whereas in the mature CNS, GABA acts as an inhibitory neurotransmitter [90, 91]. Also, during development, the establishment of synaptic connections follows a defined sequence; the GABAergic synapses are formed prior to glutamatergic ones [92]. In the adult brain, in the hippocampus, newly generated neuronal cells in the DG receive GABAergic innervations soon after their migration is completed [93]. Recent data show that the GABAergic synaptic input on neural progenitor cells in the SVZ and on nestin expressing cells – type-2 nestin-positive cells in nestin-GFP transgenic mice – in the adult DG strongly depolarizes their membranes, thereby providing an excitatory input on newly generated neuronal cells [94, 95]. The origin and mechanism of the GABA innervation on newly generated neuronal cells in the adult brain remain to be determined. The depolarization of the membrane potential by GABA during development originates from an increased chloride levels in the embryonic neural progenitor cells [96]. A similar mechanism could underlie the activity of GABA on newly generated neuronal cells in the adult brain.

During development, the depolarization of embryonic neural progenitor cells by GABA plays a role in the differentiation of granule cells [97]. In the adult brain, the GABAergic depolarization of newly generated neural progenitor cells triggers an increase of the concentration of intracellular calcium, through the activation of voltage-gated calcium channels, and the expression of NeuroD [98]. Since NeuroD is a transcription factor that is required for neuronal phenotype generation in hippocampal dentate granule neurons, the GABA depolarization on adult neural progenitor cells may also play a role in neuronal differentiation of adult progenitor cells [99]. It is further proposed that before receiving any synaptic innervations, newborn neurons may sense neuronal network activity through local ambient GABA levels [100]. Because many physiological and pathological stimulations, such as neurosteroids and epilepsy, affect GABA signaling [101-103], such mechanisms may influence the integration of new neurons in the adult brain.

Therefore, adult neurogenesis may reproduce developmental processes to integrate newly generated neuronal cells in the hippocampal network. GABA may also exert a trophic activity on neural progenitor and stem cells [104]. In the postnatal SVZ, GABA released by neuroblasts inhibits the proliferation of GFAP-progenitor cells [105]. In the SVZ and RMS, GABA released by astrocyte-like cells controls the migration of neural precursor cells [106]. Therefore, GABA may act as a paracrine factor to regulate the proliferation and migration of neural progenitor and stem cells in the adult brain.

Redefining CNS Physiopathology

A wide range of environmental stimuli and physiopathological conditions in rodents modulate neurogenesis (table 1). Environmental enrichment, exercise and learning and memory tasks stimulate neurogenesis, whereas neurogenesis decreases with aging in the DG [11, 37,107-109]. Neurogenesis is stimulated in the DG and SVZ, in the diseased brain and after CNS injuries, such as in epilepsy, strokes, traumatic brain injuries, Huntington's and Alzheimer's diseases, and decreases after stress [110-116]. New neuronal cells are also generated at the sites of degeneration where they replace some of the degenerated nerve cells, after experimental strokes [117, 118]. Cell-tracking studies revealed that new neuronal cells at the sites of degeneration originate from the SVZ. They migrate to the sites of degeneration, partially through the RMS [117, 118]. In most of these studies, neurogenesis was quantified by BrdU-labeling. Recently, Kuan et al. [119] provided evidences that hypoxia-ischemia triggers neurons to re-enter the cell cycle and resume apoptosis-associated DNA synthesis in the brain, without cell proliferation [119]. This report cautions the use of BrdU-labeling as a standard for studying neurogenesis and its regulation, and highlights the need to not only characterize BrdU-uptake, but also to show evidence of the absence of apoptotic markers, by terminal deoxynucleotidyl transferase-mediated biotinylated dUTP nick end-labeling (TUNEL) method, for example. The modulation of neurogenesis in the DG and OB in response to environmental enrichment, physiopathological conditions suggest the implication of these structures in these processes. Newly generated neuronal cells survive for extended period of time in the adult brain [14], hence environmental enrichment, physiopathological conditions may contribute to reshaping the adult CNS. Whether constitutive and induced neurogenesis recruit specific populations of progenitor cells remains to be elucidated. The ability of the CNS to regulate the generation of new neuronal cells may be use to promote brain repair in the diseased brain, and after CNS injury [120].

Investigators have reported that newly generated neuronal cells in the hippocampus are involved in learning and memory, and depression [121-123]. The function of newly generated neuronal cells, particularly in learning and memory, has been challenged by other studies, reporting that increased hippocampal neurogenesis has been observed without improvement of learning and memory performances, that learning enhances the survival of new neurons beyond the time when the hippocampus is required for memory, and that non-specific effects of treatments aiming at inhibiting adult neurogenesis have yet to be ruled out [124-126].

Table I. Modulation of neurogenesis in the adult brain

Stimuli	Area	Modulation	References
Environmental enrichment	DG	increase	37
Exercise	DG	increase	107, 109
Learning and memory tasks	DG	increase	108
Aging	DG	decrease	11
Epilepsy	DG	increase	110
Stroke	DG	increase	112
Traumatic brain injury	DG	increase	113
Huntington disease	SVZ	increase	114
Alzheimer disease	DG	increase	115
Stress	DG	decrease	111

A wide range of environmental stimuli, and physiopathological conditions modulates neurogenesis. Environmental enrichment, exercise and learning and memory tasks stimulate neurogenesis, whereas neurogenesis decreases with aging in the DG. Neurogenesis is stimulated in the DG and SVZ, in the diseased brain and after CNS injuries, such as epilepsies, strokes, traumatic brain injuries, Huntington's and Alzheimer's diseases, and decreased after stress. The modulation of neurogenesis in the DG and OB in response to environmental enrichment, physiopathological conditions suggest the implication of these structures in these processes.

Newly generated neuronal cells may also contribute to tissue homeostasis in the adult brain. The disturbance in the rate of adult neurogenesis would contribute to pathological processes in the CNS, such as in neurodegenerative diseases. However, such hypothesis remains to be investigated. Therefore, newly generated neuronal cells are involved in the physiopathology of the adult CNS, but the function and relative contribution of newly generated neuronal cells to these processes versus the neuronal cells of the pre-existing network remain to be determined. It is estimated that 0.2% of the degenerated nerve cells are replaced in the striatum after middle cerebral artery occlusion, a model of focal ischemia [117]. This low percentage of newly generated neuronal cells at the sites of injury may account for the lack of functional recovery in the injured CNS. The modulation of neurogenesis by environmental stimuli, and in physiopathological conditions may therefore contribute to CNS plasticity, and the generation of newly generated neuronal cells at the sites of injury in the diseased brain and after CNS injuries may represent a regenerative attempt by the CNS [127, 128].

Cellular Therapy in the Adult CNS

The confirmation that neurogenesis occurs in the adult brain, and that NSCs reside in the adult CNS suggest that the adult CNS may be amenable to repair. Cell therapeutic intervention may involve the stimulation and transplantation of neural progenitor and stem cells of the adult CNS. Since neural progenitor and stem cells reside throughout the adult

CNS, the stimulation of endogenous neural progenitor and stem cells locally would represent a strategy to promote regeneration in the diseased brain and after CNS injury. Particularly, since new neuronal cells that originate from the SVZ, are generated at the sites of degeneration in the diseased brain and after CNS injuries, strategies that promote regeneration and repair may focus on stimulating SVZ neurogenesis. Neural progenitor and stem cells can be isolated from the adult brain, including from human biopsies and post-mortem tissues [129, 130], providing valuable sources of tissue for cellular therapy. To support this contention, neural progenitor and stem cells have been grafted in various animals model of CNS diseases and injuries, such as Parkinson's disease and spinal cord injuries, and shown to improve functional recovery [131-137]. In these studies, the release of trophic factors by the grafted neural progenitor and stem cells, and their interaction with the injured brain are believed to play a major role in the recovery process [131-136]. In a recent study where human fetal neural progenitor and stem cells were injected after spinal cord injury in mice, the improvements in walking disappeared following treatment with diphtheria toxin, which kills only human cells – not mouse cells –, suggesting that the cells themselves are responsible for recovery. Administration of stem cells might then not simply stimulate the body to produce some healing factors, but they directly contribute to repair damage themselves [137].

Concluding Remarks

The confirmation that neurogenesis occurs in the adult brain, and that NSCs reside in the adult CNS, has open a new era in the understanding of the brain development and physiopathology, and for therapy in the adult CNS. Identifying NSCs, the function and contribution of newly generated neuronal cells to the functioning of the CNS are currently the source of intense research, and debates. The isolation and culture of neural progenitor and stem cells from the adult brain provide new opportunities for cellular therapy in the CNS. These studies will contribute to redefine our knowledge of the CNS, and bring NSC research closer to therapy.

Acknowledgments

Reproduced: Taupin P. Neurogenesis in the adult central nervous system. C R Biol (2006) 329(7): 465-75. Copyright (2006), with permission from l'Académie des sciences and Elsevier.

References

[1] S. Ramon y Cajal, Hafner: Degeneration and Regeneration of the Nervous System, New York, 1928.

[2] J. Altman, Are new neurons formed in the brains of adult mammals?, *Science* 135 (1962) 1127-1128.

[3] J. Altman, G.D. Das, Autoradiographic and histological evidence of postnatal hippocampal neurogenesis in rats, *J. Comp. Neurol.* 124 (1965) 319-336.

[4] J. Altman, Autoradiographic and histological studies of postnatal neurogenesis. IV. Cell proliferation and migration in the anterior forebrain, with special reference to persisting neurogenesis in the olfactory bulb, *J. Comp. Neurol.* 137 (1969) 433-458.

[5] M.S. Kaplan, J.W. Hinds, Neurogenesis in the adult rat: electron microscopic analysis of light radioautographs, *Science* 197 (1977) 1092-1094.

[6] S.A. Bayer, J.W. Yackel, P.S. Puri, Neurons in the rat dentate gyrus granular layer substantially increase during juvenile and adult life, *Science* 216 (1982) 890-892.

[7] H.A. Cameron, C.S. Woolley, B.S. McEwen, E. Gould, Differentiation of newly born neurons and glia in the dentate gyrus of the adult rat, *Neurosci.* 56 (1993) 337-344.

[8] F.S. Corotto, J.A. Henegar, J.A. Maruniak, Neurogenesis persists in the subependymal layer of the adult mouse brain, *Neurosci. Lett.* 149 (1993) 111-114.

[9] M.B. Luskin, Restricted proliferation and migration of postnatally generated neurons derived from the forebrain subventricular zone, *Neuron* 11 (1993) 173-189.

[10] C. Lois, A. Alvarez-Buylla, Long-distance neuronal migration in the adult mammalian brain, *Science* 264 (1994) 1145-1148.

[11] H.G. Kuhn, H. Dickinson-Anson, F.H. Gage, Neurogenesis in the dentate gyrus of the adult rat: age-related decrease of neuronal progenitor proliferation, *J. Neurosci.* 16 (1996) 2027-2033.

[12] H. Van Praag, A.F. Schinder, B.R. Christie, N. Toni, T.D. Palmer, F.H. Gage, Functional neurogenesis in the adult hippocampus, *Nature* 415 (2002) 1030-1034.

[13] E. Gould, P. Tanapat, B.S. McEwen, G. Flugge, E. Fuchs, Proliferation of granule cell precursors in the dentate gyrus of adult monkeys is diminished by stress, *Proc. Natl. Acad. Sci. USA* 95 (1998) 3168-3171.

[14] P.S. Eriksson, E. Perfilieva, T. Bjork-Eriksson, A.M. Alborn, C. Nordborg, D.A. Peterson, F.H. Gage, Neurogenesis in the adult human hippocampus, *Nat. Med.* 4 (1998) 1313-1317.

[15] D.R. Kornack, P. Rakic, The generation, migration, and differentiation of olfactory neurons in the adult primate brain, *Proc. Natl. Acad. Sci. USA* 98 (2001) 4752-4757.

[16] M.W. Miller, R.S. Nowakowski, Use of bromodeoxyuridine-immunohistochemistry to examine the proliferation, migration and time of origin of cells in the central nervous system, *Brain Res.* 457 (1988) 44-52.

[17] J.A. del Rio, E. Soriano, Immunocytochemical detection of 5'-bromodeoxyuridine incorporation in the central nervous system of the mouse, Brain Res. *Dev. Brain Res.* 49 (1989) 311-317.

[18] B.A. Reynolds, S. Weiss, Generation of neurons and astrocytes from isolated cells of the adult mammalian central nervous system, *Science* 255 (1992) 1707-1710.

[19] F.H. Gage, P.W. Coates, T.D. Palmer, H.G. Kuhn, L.J. Fisher, J.O. Suhonen, D.A. Peterson, S.T. Suhr, J. Ray, Survival and differentiation of adult neuronal progenitor cells transplanted to the adult brain, *Proc. Natl. Acad. Sci. USA* 92 (1995) 11879-11883.

[20] P. Taupin, F.H. Gage, Adult neurogenesis and neural stem cells of the central nervous system in mammals, *J. Neurosci. Res.* 69 (2002) 745-749.

[21] S.A. Goldman, M.B. Luskin, Strategies utilized by migrating neurons of the postnatal vertebrate forebrain, *Trends Neurosci.* 21 (1998) 107-114.

[22] S. Temple, A. Alvarez-Buylla, Stem cells in the adult mammalian central nervous system, *Curr. Opin. Neurobiol.* 9 (1999) 135-141.

[23] F.S. Corotto, J.A. Henegar, J.A. Maruniak, Neurogenesis persists in the subependymal layer of the adult mouse brain, *Neurosci. Lett.* 149 (1993) 111-114.

[24] V. Pencea, K.D. Bingaman, L.J. Freedman, M.B. Luskin, Neurogenesis in the subventricular zone and rostral migratory stream of the neonatal and adult primate forebrain, *Exp. Neurol.* 172 (2001) 1-16.

[25] E.C. Crosby, T. Humphrey, Studies of The Vertebrate Telencephalon. I. The nuclear configuration of the olfactory and accessory olfactory formations and of the nucleus olfactorius anterior of certain reptiles, birds, and mammals, *J. Comp. Neurol.* 71 (1939) 121-213.

[26] L. Seress, Morphological variability and developmental aspects of monkey and human granule cells: differences between the rodent and primate dentate gyrus, *Epilepsy Res. Suppl.* 7 (1992) 3-28.

[27] C. Lois, J.M. Garcia-Verdugo, A. Alvarez-Buylla, Chain migration of neuronal precursors, *Science* 271 (1996) 978-981.

[28] F. Doetsch, A. Alvarez-Buylla, Network of tangential pathways for neuronal migration in adult mammalian brain. *Proc. Natl. Acad. Sci. USA* 93 (1996) 14895-14900.

[29] P. Rousselot, C. Lois, A. Alvarez-Buylla, Embryonic (PSA) N-CAM reveals chains of migrating neuroblasts between the lateral ventricle and the olfactory bulb of adult mice, *J. Comp. Neurol.* 351 (1995) 51-61.

[30] H. Wichterle, J.M. Garcia-Verdugo, A. Alvarez-Buylla, Direct evidence for homotypic, glia-independent neuronal migration, *Neuron* 18 (1997) 779-791.

[31] K. Sawamoto, H. Wichterle, O. Gonzalez-Perez, J.A. Cholfin, M. Yamada, N. Spassky, N.S. Murcia, J.M. Garcia-Verdugo, O. Marin, J.L. Rubenstein, M. Tessier-Lavigne, H. Okano, A. Alvarez-Buylla, New neurons follow the flow of cerebrospinal fluid in the adult brain, *Science* 311 (2006) 629-632.

[32] B.B. Stanfield, J.E. Trice, Evidence that granule cells generated in the dentate gyrus of adult rats extend axonal projections, *Exp. Brain Res.* 72 (1988) 399-406.

[33] E.A. Markakis, F.H. Gage, Adult-generated neurons in the dentate gyrus send axonal projections to field CA3 and are surrounded by synaptic vesicles, *J. Comp. Neurol.* 406 (1999) 449-460.

[34] E. Gould, A.J. Reeves, M. Fallah, P. Tanapat, C.G. Gross, E. Fuchs, Hippocampal neurogenesis in adult Old World primates, *Proc. Natl. Acad. Sci. USA* 96 (1999) 5263-5267.

[35] D.R. Kornack, P. Rakic, Continuation of neurogenesis in the hippocampus of the adult macaque monkey. *Proc. Natl. Acad. Sci. USA* 96 (1999) 5768-5773.

[36] N.B. Hastings, E. Gould, Rapid extension of axons into the CA3 region by adult-generated granule cells, *J. Comp. Neurol.* 413 (1999) 146-154. Erratum in: *J. Comp. Neurol.* 415 (1999) 144.

[37] G. Kempermann, H.G. Kuhn, F.H. Gage, More hippocampal neurons in adult mice living in an enriched environment, Nature 386 (1997) 493-495.

[38] H.A. Cameron, R.D. McKay, Adult neurogenesis produces a large pool of new granule cells in the dentate gyrus, *J. Comp. Neurol.* 435 (2001) 406-417.

[39] R. Rietze, P. Poulin, S. Weiss, Mitotically active cells that generate neurons and astrocytes are present in multiple regions of the adult mouse hippocampus, *J. Comp. Neurol.* 424 (2000) 397-408.

[40] E. Gould, N. Vail, M. Wagers, C.G. Gross, Adult-generated hippocampal and neocortical neurons in macaques have a transient existence, *Proc. Natl. Acad. Sci. USA* 98 (2001) 10910-10917.

[41] E. Gould, A.J. Reeves, M.S. Graziano, C.G. Gross, Neurogenesis in the neocortex of adult primates, *Science* 286 (1999) 548-552.

[42] A. Bedard, M. Cossette, M. Levesque, A. Parent, Proliferating cells can differentiate into neurons in the striatum of normal adult monkey, *Neurosci. Lett.* 328 (2002) 213-216.

[43] P.J. Bernier, A. Bedard, J. Vinet, M. Levesque, A. Parent, Newly generated neurons in the amygdala and adjoining cortex of adult primates, *Proc. Natl. Acad. Sci. USA* 99 (2002) 11464-11469.

[44] M. Zhao, S. Momma, K. Delfani, M. Carlen, R.M. Cassidy, C.B. Johansson, H. Brismar, O. Shupliakov, J. Frisen, A.M. Janson, Evidence for neurogenesis in the adult mammalian substantia nigra, *Proc. Natl. Acad. Sci. USA* 100 (2003) 7925-7930.

[45] Y. Xu, N. Tamamaki, T. Noda, K. Kimura, Y. Itokazu, N. Matsumoto, M. Dezawa, C. Ide, Neurogenesis in the ependymal layer of the adult rat 3rd ventricle, *Exp. Neurol.* 192 (2005) 251-264.

[46] F. Luzzati, S. De Marchis, A. Fasolo, P. Peretto, Neurogenesis in the caudate nucleus of the adult rabbit, *J. Neurosci.* 26 (2006) 609-621.

[47] N.U. Takemura, Evidence for neurogenesis within the white matter beneath the temporal neocortex of the adult rat brain, *Neurosci.* 134 (2005) 121-132.

[48] D.R. Kornack, P. Rakic, Cell proliferation without neurogenesis in adult primate neocortex, *Science* 294 (2001) 2127-2130.

[49] D.C. Lie, G. Dziewczapolski, A.R. Willhoite, B.K. Kaspar, C.W. Shults, F.H. Gage, The adult substantia nigra contains progenitor cells with neurogenic potential, *J. Neurosci.* 22 (2002) 6639-6649.

[50] H. Frielingsdorf, K. Schwarz, P. Brundin, P. Mohapel, No evidence for new dopaminergic neurons in the adult mammalian substantia nigra, *Proc. Natl. Acad. Sci. USA* 101 (2004) 10177-10182.

[51] N. Sanai, A.D. Tramontin, A. Quinones-Hinojosa, N.M. Barbaro, N. Gupta, S. Kunwar, M.T. Lawton, M.W. McDermott, A.T. Parsa, J.M. Garcia-Verdugo, M.S. Berger, A. Alvarez-Buylla, Unique astrocyte ribbon in adult human brain contains neural stem cells but lacks chain migration, *Nature* 427 (2004) 740-744.

[52] P. Rakic, Limits of neurogenesis in primates, *Science* 227 (1985) 1054-1056.

[53] P. Rakic, Young neurons for old brains? *Nat. Neurosci.* 1 (1998) 645-647.

[54] F.H. Gage, Mammalian neural stem cells, *Science* 287 (2000) 1433-1438.

[55] U. Lendahl, L.B. Zimmerman, R.D. McKay, CNS stem cells express a new class of intermediate filament protein, *Cell* 60 (1990) 585-595.

[56] P.J. Yaworsky, C. Kappen, Heterogeneity of neural progenitor cells revealed by enhancers in the nestin gene, *Dev. Biol.* 205 (1999) 309-321.

[57] V.I. Chin, P. Taupin, S. Sanga, J. Scheel, F.H. Gage, S.N. Bhatia, Microfabricated platform for studying stem cell fates, *Biotechnol. Bioeng.* 88 (2004) 399-415.

[58] T. Tohyama, V.M. Lee, L.B. Rorke, M. Marvin, R.D. McKay, J.Q. Trojanowski, Nestin expression in embryonic human neuroepithelium and in human neuroepithelial tumor cells, *Lab. Invest.* 66 (1992) 303-313.

[59] S.R. Clarke, A.K. Shetty, J.L. Bradley, D.A. Turner, Reactive astrocytes express the embryonic intermediate neurofilament nestin, *Neuroreport* 5 (1994) 1885-1888.

[60] S. Sakakibara, T. Imai, K. Hamaguchi, M. Okabe, J. Aruga, K. Nakajima, D. Yasutomi, T. Nagata, Y. Kurihara, S. Uesugi, T. Miyata, M. Ogawa, K. Mikoshiba, H. Okano, Mouse-Musashi-1, a neural RNA-binding protein highly enriched in the mammalian CNS stem cell, *Dev. Biol.* 176 (1996) 230-242.

[61] M.V. Zappone, R. Galli, R. Catena, N. Meani, S. De Biasi, E. Mattei, C. Tiveron, A.L. Vescovi, R. Lovell-Badge, S. Ottolenghi, S.K. Nicolis, Sox2 regulatory sequences direct expression of a (beta)-geo transgene to telencephalic neural stem cells and precursors of the mouse embryo, revealing regionalization of gene expression in CNS stem cells, *Development* 127 (2000) 2367-2382.

[62] Y. Kaneko, S. Sakakibara, T. Imai, A. Suzuki, Y. Nakamura, K. Sawamoto, Y. Ogawa, Y. Toyama, T. Miyata, H. Okano, Musashi1: an evolutionally conserved marker for CNS progenitor cells including neural stem cells, *Dev. Neurosci.* 22 (2000) 139-153.

[63] T. Okuda, K. Tagawa, M.L. Qi, M. Hoshio, H. Ueda, H. Kawano, I. Kanazawa, M. Muramatsu, H. Okazawa, Oct-3/4 repression accelerates differentiation of neural progenitor cells in vitro and in vivo, *Brain Res. Mol. Brain Res.* 132 (2004) 18-30.

[64] M. Komitova, P.S. Eriksson, Sox-2 is expressed by neural progenitors and astroglia in the adult rat brain. *Neurosci. Lett.* 369 (2004) 24-27.

[65] D.H. Geschwind, J. Ou, M.C. Easterday, J.D. Dougherty, R.L. Jackson, Z. Chen, H. Antoine, A. Terskikh, I.L. Weissman, S.F. Nelson, H.I. Kornblum, A genetic analysis of neural progenitor differentiation, *Neuron* 29 (2001) 325-339.

[66] H.I. Kornblum, D.H. Geschwind DH, Molecular markers in CNS stem cell research: hitting a moving target, *Nat. Rev. Neurosci.* 2 (2001) 843-846.

[67] O.N. Suslov, V.G. Kukekov, T.N. Ignatova, D.A. Steindler, Neural stem cell heterogeneity demonstrated by molecular phenotyping of clonal neurospheres, *Proc. Natl. Acad. Sci. USA* 99 (2002) 14506-14511.

[68] P. Taupin, J. Ray, W.H. Fischer, S.T. Suhr, K. Hakansson, A. Grubb, F.H. Gage, FGF-2-responsive neural stem cell proliferation requires CCg, a novel autocrine/paracrine cofactor, *Neuron* 28 (2000) 385-397.

[69] B.A. Reynolds, R.L. Rietze, Neural stem cells and neurospheres-re-evaluating the relationship, *Nat. Methods* 2 (2005) 333-336.

[70] A. Gritti, E.A. Parati, L. Cova, P. Frolichsthal, R. Galli, E. Wanke, L. Faravelli, D.J. Morassutti, F. Roisen, D.D. Nickel, A.L. Vescovi, Multipotential stem cells from the adult mouse brain proliferate and self-renew in response to basic fibroblast growth factor. *J. Neurosci.* 16 (1996) 1091-1100.

[71] T.D. Palmer, J. Takahashi, F.H. Gage, The adult rat hippocampus contains primordial neural stem cells, *Mol. Cell. Neurosci.* 8 (1997) 389-404.

[72] R.M. Seaberg, D. van der Kooy, Adult rodent neurogenic regions: the ventricular subependyma contains neural stem cells, but the dentate gyrus contains restricted progenitors, *J. Neurosci.* 22 (2002) 1784-1793.

[73] N.D. Bull, P.F. Bartlett, The adult mouse hippocampal progenitor is neurogenic but not a stem cell, *J. Neurosci.* 25 (2005) 10815-10821.

[74] F. Doetsch, I. Caille, D.A. Lim, J.M. Garcia-Verdugo, A. Alvarez-Buylla, Subventricular zone astrocytes are neural stem cells in the adult mammalian brain, *Cell* 97 (1999) 703-716.

[75] C.B. Johansson, S. Momma, D.L. Clarke, M. Risling, U. Lendahl, J. Frisen, Identification of a neural stem cell in the adult mammalian central nervous system, *Cell* 96 (1999) 25-34.

[76] B.J. Chiasson, V. Tropepe, C.M. Morshead, D. van der Kooy, Adult mammalian forebrain ependymal and subependymal cells demonstrate proliferative potential, but only subependymal cells have neural stem cell characteristics, *J. Neurosci.* 19 (1999) 4462-4471.

[77] E.D. Laywell, P. Rakic, V.G. Kukekov, E.C. Holland, D.A. Steindler, Identification of a multipotent astrocytic stem cell in the immature and adult mouse brain, *Proc. Natl. Acad. Sci. USA* 97 (2000) 13883-13888.

[78] B. Seri, J.M. Garcia-Verdugo, B.S. McEwen, A. Alvarez-Buylla, Astrocytes give rise to new neurons in the adult mammalian hippocampus, *J. Neurosci.* 21 (2001) 7153-7160.

[79] V. Filippov, G. Kronenberg, T. Pivneva, K. Reuter, B. Steiner, L.P. Wang, M. Yamaguchi, H. Kettenmann, G. Kempermann, Subpopulation of nestin-expressing progenitor cells in the adult murine hippocampus shows electrophysiological and morphological characteristics of astrocytes, *Mol. Cell. Neurosci.* 23 (2003) 373-382.

[80] F. Doetsch, J.M. Garcia-Verdugo, A. Alvarez-Buylla, Cellular composition and three-dimensional organization of the subventricular germinal zone in the adult mammalian brain, *J. Neurosci.* 17 (1997) 5046-5061.

[81] J.M. Garcia-Verdugo, F. Doetsch, H. Wichterle, D.A. Lim, A. Alvarez-Buylla, Architecture and cell types of the adult subventricular zone: in search of the stem cells, *J. Neurobiol.* 36 (1998) 234-248.

[82] C.M. Morshead, B.A. Reynolds, C.G. Craig, M.W. McBurney, W.A. Staines, D. Morassutti, S. Weiss, D. van der Kooy, Neural stem cells in the adult mammalian forebrain: a relatively quiescent subpopulation of subependymal cells, *Neuron* 13 (1994) 1071-1082.

[83] F. Doetsch, J.M. Garcia-Verdugo, A. Alvarez-Buylla, Regeneration of a germinal layer in the adult mammalian brain, *Proc. Natl. Acad. Sci. USA* 96 (1999) 11619-11624.

[84] B. Seri, J.M. Garcia-Verdugo, L. Collado-Morente, B.S. McEwen, A. Alvarez-Buylla, Cell types, lineage, and architecture of the germinal zone in the adult dentate gyrus, *J. Comp. Neurol.* 478 (2004) 359-378. Erratum in: *J. Comp. Neurol.* 480 (2004) 427.

[85] V.L. Savchenko, J.A. McKanna, I.R. Nikonenko, G.G. Skibo, Microglia and astrocytes in the adult rat brain: comparative immunocytochemical analysis demonstrates the efficacy of lipocortin 1 immunoreactivity, *Neurosci.* 96 (2000) 195-203.

[86] C.M. Morshead, A.D. Garcia, M.V. Sofroniew, D. van Der Kooy, The ablation of glial fibrillary acidic protein-positive cells from the adult central nervous system results in the loss of forebrain neural stem cells but not retinal stem cells, *Eur. J. Neurosci.* 18 (2003) 76-84.

[87] T. Imura, H.I. Kornblum, M.V. Sofroniew, The predominant neural stem cell isolated from postnatal and adult forebrain but not early embryonic forebrain expresses GFAP, *J. Neurosci.* 23 (2003) 2824-2832.

[88] A.D. Garcia, N.B. Doan, T. Imura, T.G. Bush, M.V. Sofroniew, GFAP-expressing progenitors are the principal source of constitutive neurogenesis in adult mouse forebrain, *Nat. Neurosci.* 7 (2004) 1233-1241.

[89] H. Song, C.F. Stevens, F.H. Gage, Astroglia induce neurogenesis from adult neural stem cells, *Nature* 417 (2002) 39-44.

[90] Y. Ben-Ari, Excitatory actions of gaba during development: the nature of the nurture, *Nat. Rev. Neurosci.* 3 (2002) 728-739.

[91] D.F. Owens, A.R. Kriegstein, Is there more to GABA than synaptic inhibition? *Nat. Rev. Neurosci.* 3 (2002) 715-727.

[92] P. Davies, B. Anderton, J. Kirsch, A. Konnerth, R. Nitsch, M. Sheetz, First one in, last one out: the role of gabaergic transmission in generation and degeneration, *Prog. Neurobiol.* 55 (1998) 651–658.

[93] L.P. Wang, G. Kempermann, H. Kettenmann, A subpopulation of precursor cells in the mouse dentate gyrus receives synaptic GABAergic input, *Mol. Cell. Neurosci.* 29 (2005) 181-189.

[94] Y. Tozuka, S. Fukuda, T. Namba, T. Seki, T. Hisatsune, GABAergic excitation promotes neuronal differentiation in adult hippocampal progenitor cells, *Neuron* 47 (2005) 803-815.

[95] D.D. Wang, D.D. Krueger, A. Bordey, GABA depolarizes neuronal progenitors of the postnatal subventricular zone via GABAA receptor activation, *J. Physiol.* 550 (2003) 785-800.

[96] J.J. LoTurco, D.F. Owens, M.J. Heath, M.B. Davis, A.R. Kriegstein, GABA and glutamate depolarize cortical progenitor cells and inhibit DNA synthesis, *Neuron* 15 (1995) 1287-1298.

[97] L.C. Katz, C.J. Shatz, Synaptic activity and the construction of cortical circuits, *Science* 274 (1996) 1133-1138.

[98] M.H. Schwab, A. Bartholomae, B. Heimrich, D. Feldmeyer, S. Druffel-Augustin, S. Goebbels, F.J. Naya, S. Zhao, M. Frotscher, M.J. Tsai, K.A. Nave, Neuronal basic helix-loop-helix proteins (NEX and BETA2/Neuro D) regulate terminal granule cell differentiation in the hippocampus, *J. Neurosci.* 20 (2000) 3714-3724. Erratum in: *J. Neurosci.* 20 (2000) 8227.

[99] K. Deisseroth, S. Singla, H. Toda, M. Monje, T.D. Palmer, R.C. Malenka, Excitation-neurogenesis coupling in adult neural stem/progenitor cells, *Neuron* 42 (2004) 535-552.

[100] S. Ge, E.L. Goh, K.A. Sailor, Y. Kitabatake, G.L. Ming, H. Song, GABA regulates synaptic integration of newly generated neurons in the adult brain, *Nature* 439 (2006) 589-593.

[101] A. Semyanov, M.C. Walker, D.M. Kullmann, R.A. Silver, Tonically active GABA A receptors: modulating gain and maintaining the tone, *Trends Neurosci.* 27 (2004) 262-269.

[102] M. Farrant, Z. Nusser, Variations on an inhibitory theme: phasic and tonic activation of GABA(A) receptors, *Nat. Rev. Neurosci.* 6 (2005) 215-229.

[103] D. Belelli, J.J. Lambert, Neurosteroids: endogenous regulators of the GABA(A) receptor, *Nat. Rev. Neurosci.* 6 (2005) 565-575.

[104] S.A. Lipton, S.B. Kater, Neurotransmitter regulation of neuronal outgrowth, plasticity and survival, *Trends Neurosci.* 12 (1989) 265-270.

[105] X. Liu, Q. Wang, T.F. Haydar, A. Bordey, Nonsynaptic GABA signaling in postnatal subventricular zone controls proliferation of GFAP-expressing progenitors, *Nat. Neurosci.* 8 (2005) 1179-1187.

[106] A.J. Bolteus, A. Bordey, GABA release and uptake regulate neuronal precursor migration in the postnatal subventricular zone, *J. Neurosci.* 24 (2004) 7623-7631.

[107] H. van Praag, G. Kempermann, F.H. Gage, Running increases cell proliferation and neurogenesis in the adult mouse dentate gyrus, *Nat. Neurosci.* 2 (1999) 266-270.

[108] E. Gould, A. Beylin, P. Tanapat, A. Reeves, T.J. Shors, Learning enhances adult neurogenesis in the hippocampal formation, *Nat. Neurosci.* 2 (1999) 260-265.

[109] S.M. Ra, H. Kim, M.H. Jang, M.C. Shin, T.H. Lee, B.V. Lim, C.J. Kim, E.H. Kim, K.M. Kim, S.S. Kim, Treadmill running and swimming increase cell proliferation in the hippocampal dentate gyrus of rats, *Neurosci. Lett.* 333 (2002) 123-126.

[110] J.M. Parent, T.W. Yu, R.T. Leibowitz, D.H. Geschwind, R.S. Sloviter, D.H. Lowenstein DH, Dentate granule cell neurogenesis is increased by seizures and contributes to aberrant network reorganization in the adult rat hippocampus, *J. Neurosci.* 17 (1997) 3727-3738.

[111] E. Gould, P. Tanapat, B.S. McEwen, G. Flugge, E. Fuchs, Proliferation of granule cell precursors in the dentate gyrus of adult monkeys is diminished by stress, *Proc. Natl. Acad. Sci. USA* 95 (1998) 3168-3171.

[112] J. Liu, K. Solway, R.O. Messing, F.R. Sharp, Increased neurogenesis in the dentate gyrus after transient global ischemia in gerbils, *J. Neurosci.* 18 (1998) 7768-7778.

[113] P.K. Dash, S.A. Mach, A.N. Moore, Enhanced neurogenesis in the rodent hippocampus following traumatic brain injury, *J. Neurosci. Res.* 63 (2001) 313-319.

[114] M.A. Curtis, E.B. Penney, A.G. Pearson, W.M. van Roon-Mom, N.J. Butterworth, M. Dragunow, B. Connor, R.L. Faull, Increased cell proliferation and neurogenesis in the adult human Huntington's disease brain, *Proc. Natl. Acad. Sci. USA* 100 (2003) 9023-9027.

[115] K. Jin, A.L. Peel, X.O. Mao, L. Xie, B.A. Cottrell, D.C. Henshall, D.A. Greenberg, Increased hippocampal neurogenesis in Alzheimer's disease, *Proc. Natl. Acad. Sci. USA* 101 (2004) 343-347.

[116] P. Taupin, Neurogenesis in the pathologies of the nervous system, *Med. Sci.* (Paris) 21 (2005) 711-714.

[117] A. Arvidsson, T. Collin, D. Kirik, Z. Kokaia, O. Lindvall, Neuronal replacement from endogenous precursors in the adult brain after stroke, *Nat. Med.* 8 (2002) 963-970.
[118] K. Jin, Y. Sun, L. Xie, A. Peel, X.O. Mao, S. Batteur, D.A. Greenberg, Directed migration of neuronal precursors into the ischemic cerebral cortex and striatum, *Mol. Cell. Neurosci.* 24 (2003) 171-189.
[119] C.Y. Kuan, A.J. Schloemer, A. Lu, K.A. Burns, W.L. Weng, M.T. Williams, K.L. Strauss, C.V. Vorhees, R.A. Flavell, R.J. Davis, F.R. Sharp, P. Rakic, Hypoxia-ischemia induces DNA synthesis without cell proliferation in dying neurons in adult rodent brain, *J. Neurosci.* 24 (2004) 10763-10772.
[120] P. Taupin, Adult neurogenesis in the mammalian central nervous system: functionality and potential clinical interest, *Med. Sci. Monit.* 11 (2005) RA247-252.
[121] T.J. Shors, G. Miesegaes, A. Beylin, M. Zhao, T. Rydel, E. Gould, Neurogenesis in the adult is involved in the formation of trace memories, *Nature* 410 (2001) 372-376. Erratum in: *Nature* 414 (2001) 938.
[122] L. Santarelli, M. Saxe, C. Gross, A. Surget, F. Battaglia, S. Dulawa, N. Weisstaub, J. Lee, R. Duman, O. Arancio, C. Belzung, R. Hen, Requirement of hippocampal neurogenesis for the behavioral effects of antidepressants, *Science* 301 (2003) 805-809.
[123] J.S. Snyder, N.S. Hong, R.J. McDonald, J.M. Wojtowicz, A role for adult neurogenesis in spatial long-term memory, *Neurosci.* 130 (2005) 843-852.
[124] J.S. Rhodes, H. van Praag, S. Jeffrey, I. Girard, G.S. Mitchell, T. Jr Garland, F.H. Gage, Exercise increases hippocampal neurogenesis to high levels but does not improve spatial learning in mice bred for increased voluntary wheel running, *Behav. Neurosci.* 117 (2003) 1006-1016. Erratum in: *Behav. Neurosci.* 118 (2003) 305.
[125] B. Leuner, S. Mendolia-Loffredo, Y. Kozorovitskiy, D. Samburg, E. Gould, T.J. Shors, Learning enhances the survival of new neurons beyond the time when the hippocampus is required for memory, *J. Neurosci.* 24 (2004) 7477-7481.
[126] D. Dupret, M.F. Montaron, E. Drapeau, C. Aurousseau, M. Le Moal, P.V. Piazza, D.N. Abrous, Methylazoxymethanol acetate does not fully block cell genesis in the young and aged dentate gyrus, *Eur. J. Neurosci.* 22 (2005) 778-783.
[127] P. Taupin, Adult neurogenesis and neuroplasticity, Restor. *Neurol. Neurosci.* 24 (2006) 9-15.
[128] P. Taupin, Stroke-induced neurogenesis: physiopathology and mechanisms, *Curr. Neurovasc. Res.* 3 (2006) 67-72.
[129] N.S. Roy, S. Wang, L. Jiang, J. Kang, A. Benraiss, C. Harrison-Restelli, R.A. Fraser, W.T. Couldwell, A. Kawaguchi, H. Okano, M. Nedergaard, S.A. Goldman, In vitro neurogenesis by progenitor cells isolated from the adult human hippocampus, *Nat. Med.* 6 (2000) 271-277.
[130] T.D. Palmer, P.H. Schwartz, P. Taupin, B. Kaspar, S.A. Stein, F.H. Gage, Cell culture. Progenitor cells from human brain after death, *Nature* 411 (2001) 42-43.
[131] J. Ourednik, V. Ourednik, W.P. Lynch, M. Schachner, E.Y. Snyder, Neural stem cells display an inherent mechanism for rescuing dysfunctional neurons, *Nat. Biotechnol.* 20 (2002) 1103-1110.

[132] P. Lu, L.L. Jones, E.Y. Snyder, M.H. Tuszynski, Neural stem cells constitutively secrete neurotrophic factors and promote extensive host axonal growth after spinal cord injury, *Exp. Neurol.* 181 (2003) 115-129.

[133] J. Yan, A.M. Welsh, S.H. Bora, E.Y. Snyder, V.E. Koliatsos, Differentiation and tropic/trophic effects of exogenous neural precursors in the adult spinal cord, *J. Comp. Neurol.* 480 (2004) 101-114.

[134] K.B. Bjugstad, D.E. Jr Redmond, Y.D. Teng, J.D. Elsworth, R.H. Roth, B.C. Blanchard, E.Y. Snyder, J.R. Jr Sladek, Neural stem cells implanted into MPTP-treated monkeys increase the size of endogenous tyrosine hydroxylase-positive cells found in the striatum: a return to control measures, *Cell. Transplant.* 14 (2005) 183-192.

[135] K. Pfeifer, M. Vroemen, A. Blesch, N. Weidner, Adult neural progenitor cells provide a permissive guiding substrate for corticospinal axon growth following spinal cord injury, *Eur. J. Neurosci.* 20 (2004) 1695-1704.

[136] K.I. Park, Y.D. Teng, E.Y. Snyder, The injured brain interacts reciprocally with neural stem cells supported by scaffolds to reconstitute lost tissue, *Nat. Biotechnol.* 20 (2002) 1111-1117.

[137] B.J. Cummings, N. Uchida, S.J. Tamaki, D.L. Salazar, M. Hooshmand, R. Summer, F.H. Gage, A.J. Anderson, Human neural stem cells differentiate and promote locomotor recovery in spinal cord-injured mice, *Proc. Natl. Acad. Sci. USA* 102 (2005) 14069-14074.

Chapter II

BrdU Immunohistochemistry for Studying Adult Neurogenesis: Paradigms, Pitfalls, Limitations and Validation

Abstract

Bromodeoxyuridine (BrdU) is a thymidine analog that incorporates DNA of dividing cells during the S-phase of the cell cycle. As such, BrdU is used for birthdating and monitoring cell proliferation. BrdU immunohistochemistry has been instrumental for the study of the development of the nervous system, and to confirm that neurogenesis occurs in the adult mammalian brain, including in human. However, the use of BrdU for studying neurogenesis is not without pitfalls and limitations. BrdU is a toxic and mutagenic substance. It triggers cell death, the formation of teratomas, alters DNA stability, lengthens the cell cycle, and has mitogenic, transcriptional and translational effects on cells that incorporate it. All of which have profound consequences on neurogenesis. BrdU is not a marker of the S-phase of the cell cycle. As a thymidine analog, it is a marker of DNA synthesis. Therefore studying neurogenesis with BrdU requires distinguishing cell proliferation and neurogenesis from other events involving DNA synthesis, like DNA repair, abortive cell cycle re-entry and gene duplication. BrdU labeling is currently the most used technique for studying adult neurogenesis *in situ*. However in many instances, appropriate controls have been overlooked and events reported as the generation of new neuronal cells in the adult brain misinterpreted, which makes BrdU labeling one of the most misused techniques in neuroscience.

1. Introduction

For decades, tritiated ([3H]) thymidine autoradiography has been used to study cell proliferation *in situ*, and to determine the time of origin, migration, lineage and fate of neuronal cells in the developing central nervous system (CNS) (Sidman et al., 1959; Angevine, 1965; Rakic, 1974; Schlessinger et al., 1975; Crespo et al., 1986; Altman and

Bayer, 1990). [3H]-thymidine autoradiography provided also the first evidences that neurogenesis occurs in discrete areas of the adult mammalian brain -in rodent-, the subventricular zone (SVZ) and the dentate gyrus (DG) of the hippocampus (Altman and Das, 1965; Altman, 1969). The use of a radiolabeled substrate and the time-consuming process involved in [3H]-thymidine autoradiography mandated the development of new strategies to study cell proliferation, and neurogenesis. BrdU is a halopyrimidine used as an antiviral and antineoplastic agent (Freese et al., 1994; Begg et al., 2000). BrdU labeling was developed as an alternative approach for determining the proliferative index of tumors (Hoshino et al., 1989; Struikmans et al., 1997). BrdU can be revealed by immunohistochemistry using a monoclonal antibody (Gratzner, 1982). BrdU immunohistochemistry offers several advantages over [3H]-thymidine autoradiography. It allows faster studies without handling radiolabeled material, and the detection of labeled cells throughout the relatively thick tissue sections required for stereological studies of the brain (West et al., 1991). This has lead to the wide use of BrdU for studying adult neurogenesis. Currently BrdU labeling is the most used technique for studying adult neurogenesis *in situ*. However, the use of BrdU and thymidine analogs is not without pitfalls and limitations (Nowakowski and Hayes, 2001; Gould and Gross, 2002; Rakic, 2002). BrdU is a toxic and mutagenic substance that induces many side effects. BrdU is not a marker for cell proliferation; it is marker of DNA synthesis. This has profound consequences on the study and analysis of adult neurogenesis, as it may lead to false results and misinterpretation if appropriate controls are not performed. In this article, we will review the paradigms, pitfalls, limitations and validation of BrdU immunohistochemistry for studying adult neurogenesis.

2. Adult Neurogenesis

2.1. In the Mammalian Brain

It is now well established that neurogenesis occurs throughout adulthood in the mammalian brain (Gage, 2000), including in human (Eriksson et al., 1998). Neurogenesis occurs primarily in two areas of the mammalian brain, the DG and the SVZ (Taupin and Gage, 2002). In the DG, newly generated neuronal cells in the subgranular zone (SGZ), a layer beneath the granular layer, migrate to the granular layer where they differentiate into neuronal cells of the granular layer and extend axonal projections to the CA3 area of the Ammon's horn (Cameron et al., 1993; Stanfield and Trice, 1988; Markakis and Gage, 1999). Newborn cells in the anterior part of the SVZ migrate through the rostro-migratory stream (RMS) to the olfactory bulb (OB), where they differentiate into interneurons, granule and periglomerular neurons (Lois and Alvarez-Buylla, 1994). Newly generated neuronal cells in the DG and OB establish synaptic contacts, and functional connections with neighboring cells (Markakis and Gage, 1999; van Praag et al., 2002; Carlen et al., 2002; Belluzzi et al., 2003).

The first evidences that neurogenesis occurs in the adult rodent brain were reported by Altman and Das (1965) and Altman (1969), using [3H]-thymidine autoradiography. These studies were substantiated by others (Kaplan and Hinds, 1977; Bayer et al., 1982), and their findings extended to the identification of the mode of migration and origin of newly generated neuronal cells. Newly generated neuronal cells of the SGZ migrating to the granular layer are associated with radial glial-like cells (Seki and Arai, 1999), whereas newly generated neuronal cells in the SVZ migrate by chain migration (Rousselot et al., 1995; Lois et al., 1996), using each other as the migratory substrate, without axonal or radial guides (Doetsch and Alvarez-Buylla, 1996; Wichterle et al., 1997). Both in the SGZ and SVZ, a glial origin for newly generated neuronal cells has been proposed (Doetsch et al., 1999; Seri et al., 2001). The occurrence of neurogenesis in adult rodent was confirmed by BrdU and retroviral labeling studies (Seki and Arai, 1993; Corotto et al., 1993; Kuhn et al., 1996; Praag et al., 2002; Yamada et al., 2004). In primate, an initial report, using [3H]-thymidine labeling, showing that neurogenesis does not occur in the adult brain of non-human primates (rhesus monkey) (Rakic, 1985) was challenged by subsequent studies using BrdU labeling. In macaque monkey and marmoset, neurogenesis occurs in the adult DG and SVZ, with similar features than in rodent (Gould et al., 1998, 1999a,b; Kornack and Rakic, 1999; 2001a; Pencea et al., 2001). Eriksson et al. (1998) using BrdU labeling, reported the first evidence that neurogenesis occurs in the adult human brain. Newborn neuronal cells immuno-positive for BrdU and neuronal nuclear antigen (NeuN), a mature neuronal marker, were identified in the adult DG of postmortem tissue samples obtained from cancer patients who had been administered with the drug as part of their treatment. In another study, Sanai et al. (2004) using immunohistofluorescence labeling for markers of the cell cycle, like Ki-67, found no evidence of a migrating pathway of neuroblasts from the SVZ to the OB of tissues obtained from neurosurgical resections and autopsied brains. The authors concluded that neurons in the adult human OB are not replaced, or that precursors may migrate as individual cells rather than as migrating chains in human SVZ (Sanai et al., 2004). Alternatively, neuroblasts from the SVZ may migrate to the human OB, through a different path, yet to be identified (Taupin, 2006). Bedard and Parent (2004), using immunohistofluorescence labeling for cell cycle markers, like Ki-67, and doublecortin, a marker of immature neurons, showed that new neuronal cells are generated in the adult human OB. The origin of these new cells in the human OB remains to be determined.

Neurogenesis has been reported to occur in other areas of the adult brain in certain species, albeit at lower level. Neurogenesis has been reported to occur in the CA1 area of the hippocampus (Rietze et al, 2000) and substantia nigra (SN) (Zhao et al., 2003) of adult mice, and in the neocortex (Gould et al., 1999a,b, 2001), striatum (Bedard et al., 2002, 2006) and amygdala (Bernier et al., 2002) of adult non-human primates. Some of these results have been contradicted by others. Lie et al. (2002) and Frielingsdorf et al. (2004) did not report evidence for new dopaminergic neurons in the adult mammalian SN, whereas Kornack and Rakic (2001b) and Koketsu et al. (2003) reported cell proliferation without neurogenesis in adult primate neocortex. Using C14-dating of DNA, Spalding at al. (2005) reported that nerves cells in the cortex of the adult human brain are not replaced. The confirmation of neurogenesis in these latter areas of the adult mammalian brain remains to be further investigated.

2.2. Temporal and Spatial Expression of Markers of Newly Generated Neuronal Cells

Multiple labeling studies, using BrdU immunohistochemistry and confocal microscopy, reveal that newly generated neuronal cells in the adult brain express, temporally and spatially, a sequence of markers (Kuhn et al., 1996; Palmer et al., 2000). In the rat DG, in the hours following BrdU administration, newborn cells express markers of the cell cycle, like Ki-67, as the S-phase cells that have incorporated BrdU are completing their cell cycle. As early as 2 days after BrdU injection, newly generated neuronal cells express nestin, a marker of neural progenitor and stem cells (Lendahl et al., 1990). During the first week following BrdU injection, nestin- and BrdU-positive cells are detected in the granular layer. Between day 2 and day 10 following BrdU administration, newborn cells express markers for immature neuronal cells, like class III β-tubulin isotype (Tuj-1) and doublecortin (Brown et al., 2003). The number of nestin- and BrdU-positive cells decreases, whereas the proportion of immature neurons increased over the first week. Seven to 10 days after BrdU injection, newborn cells in the granular layer express mature neuronal markers, calbindin and NeuN (Kuhn et al., 1996). After one month, 70% of BrdU-positive cells express mature neuronal markers, whereas the number of immature precursors had decreased to less than 10% in the granular layer (Palmer et al., 2000). This shows that newly generated neuronal cells in the DG express progressively and sequentially molecular markers that reflect neuronal maturation, from progenitor cells in the SGZ to terminally differentiated neuronal cells of the granular layer. The maturation of newborn cells, from the proliferation of newly generated cells in the SGZ to the migration and differentiation in neuronal cells of the granular layer, takes approximately 4 weeks in the adult rat DG (Cameron et al., 1993). Newly generated neuronal cells in the granular layer send axonal projections to their target cells, as early as 4-10 days after mitosis (Hastings and Gould, 1999).

2.3. Quantitative Studies

Quantitative studies revealed that in adult rodent as many as 9,000 new neuronal cells are generated per day in the DG, contributing to about 3.3% per month or 0.1% per day of the granule cell population (Cameron and McKay, 2001; Kempermann et al., 1997). In the SVZ, as many as 30,000 newly generated neuronal progenitor cells converge per day to the RMS, to feed persisting neurogenesis in the OB (Alvarez-Buylla et al., 2001). In the OB, 65.3-76.9% of the bulbar neurons are replaced during a 6 weeks period (Kato et al., 2001). In adult macaque, at least 0.004% of neuronal cells of the granule cell layer are new neurons generated per day (Kornack and Rakic, 1999). The rate of neurogenesis in human DG was also reported to be low (Eriksson et al., 1998). These quantifications were primarily performed by BrdU labeling and confocal microscopy, by counting the number of BrdU-positive cells by stereology. Since different doses of BrdU were used in the studies, and as the permeability of the blood brain barrier (BBB) to BrdU may differ between species, it is difficult to compare the studies. These issues will be discussed in the following sections of the manuscript.

2.4. Modulation of Neurogenesis

Reports show that adult neurogenesis is modulated by a broad range of experimental and physiopathological conditions, in rodent and primate. Using BrdU labeling, Kempermann et al. (1977) reported that environmental enrichment promotes the survival of newly generated neuronal cells in the hippocampus of adult rodent. Similar studies showed that physical activity stimulates hippocampal neurogenesis (van Praag et al., 1999), and that various trophic factors, neurotransmitters and drugs modulate neurogenesis in the DG and SVZ (Craig et al., 1996; Kuhn et al., 1997; Brezun and Daszuta, 1999; Malberg et al., 2000; Jin et al., 2006). Using immunohistochemistry against markers of the cell cycle, Curtis et al. (2003) and Jin et al. (2004a) showed that neurogenesis is enhanced in the SVZ of Huntington's disease (HD), and in the hippocampus of Alzheimer's disease (AD) brains, respectively. An increase in neurogenesis in animals models of HD and AD was subsequently reported by the same groups, by means of BrdU labeling (Jin et al., 2004b; Tattersfield et al., 2004). Neurogenesis is increased in the hippocampus and SVZ in animal models of epilepsy (Parent et al., 1997), strokes (Liu et al., 1998) and traumatic brain injuries (Dash et al., 2001). In the diseased and injured brain, like in HD and after strokes, new neuronal cells migrate from the SVZ toward the site of degeneration, where they replace some of the lost nerve cells (Arvidsson et al., 2002; Jin et al., 2003; Curtis et al., 2003). An estimated 0.2% of the degenerated nerve cells are replaced in the striatum after middle cerebral artery occlusion, a model of focal ischemia (Arvidsson et al., 2002). Experimental lesions or infusion of trophic factors have also been reported to induce the generation of new neuronal cells in the neocortex (Magavi et al., 2000), the hypothalamus (Kokoeva et al., 2005), and of new corticospinal motor neurons in the spinal cord (Chen et al., 2004). Recently, Danilov et al. (2006) reported the generation of new neuronal cells in the spinal cord of an animal model of multiple sclerosis.

3. Strategies for Studying Adult Neurogenesis

Thymidine analogs incorporate DNA of dividing cells during the S-phase of the cell cycle. As such, they are used for birthdating and monitoring cell proliferation. For decades, [3H]-thymidine autoradiography has been used to study cell proliferation, and neurogenesis in the developing and adult CNS (Sidman, 1970; Kaplan and Hinds, 1977; Bayer et al., 1982; Rakic, 1985; Doetsch and Alvarez-Buylla, 1996). The use of a radiolabeled substrate and the time-consuming process involved in [3H]-thymidine autoradiography mandated the development of new strategies for studying cell proliferation, and neurogenesis.

3.1. BrdU Labeling

BrdU (5-bromo-2'-deoxyuridine, figure 1) is a halopyrimidine used therapeutically as antiviral and antineoplastic agent (Begg et al., 2000; Freese et al., 1994). BrdU can be detected by immunohistochemistry, using a monoclonal antibody directed against single stranded DNA containing BrdU (Gratzner, 1982). BrdU labeling was developed as an alternative approach for determining the proliferative index of tumors (Hoshino et al., 1989; Struikmans et al., 1997), and was introduced for studying cell proliferation in the developing nervous system by Nowakowski et al. (1989).

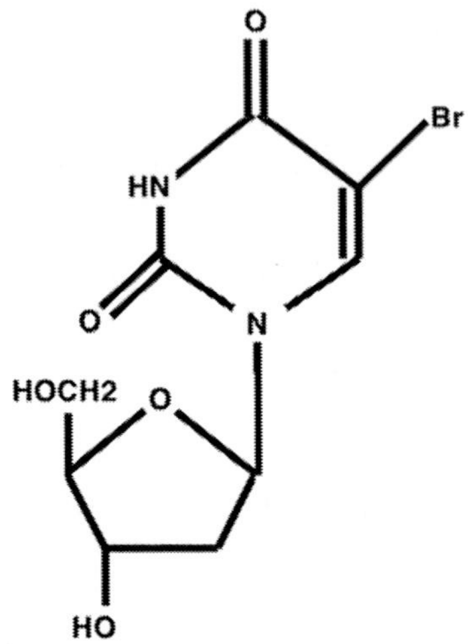

Figure 1. BrdU is a halopyrimidine. BrdU (5-bromo-2'-deoxyuridine) is a thymidine analog that incorporates DNA of dividing cells during the S-phase of the cell cycle, and is used for birthdating and monitoring cell proliferation. It is a toxic and mutagenic substance that has side effects. The integration of a bromine atom into the DNA alters its stability, increasing the risk of sister-chromatid exchanges, mutations, DNA double-strand breaks, and lengthens the cell cycle of cells that incorporate it. BrdU can be revealed by immunohistochemistry using a monoclonal antibody directed against single stranded DNA containing BrdU. BrdU immunohistochemistry has been instrumental to confirm that neurogenesis occurs in the adult mammalian brain, including in human, and is currently the most used technique for studying adult neurogenesis. However, the use of BrdU labeling is not without limitations and pitfalls. BrdU has mitogenic, transcriptional and translational effects on cells that incorporate it. BrdU is not a marker of the S-phase of the cell cycle; it is a marker of DNA synthesis. Therefore, it also labels cells undergoing DNA repair, abortive cell cycle re-entry, as a prelude to apoptosis, and gene duplication without cell division, leading to polyploidy.

3.1.1. Mode of Administration

BrdU crosses the BBB, and it may be delivered by intracerebroventricular (i.c.v.), intraperitoneal (i.p.), intravenous (i.v.) injection, or orally for studying adult neurogenesis. BrdU is metabolized through dehalogenation when integrated into the DNA. Once dehalogenated, the uracil residue would be excised from the DNA by the uracil glycosylase repair system (Hume at al., 1986). BrdU is also metabolized rapidly through dehalogenation in the plasma (the half-life of BrdU in the plasma is reported to be around 8-11 minutes in human) (Kriss et al., 1963). The concentration of thymidine analog that reaches the brain is therefore only a fraction of the administered dose. Hence, i.c.v. injection leads to higher concentration of BrdU in the brain than peripheral delivery, and is employed when a higher and local concentration of BrdU in the brain is sought (Kaplan, 1983; Zhao et al., 2003). However, i.p. or i.v. injections are the most common mode of administration of BrdU in rodent and primate for studying adult neurogenesis, as they obviate the need for surgery

(Kuhn et al., 1996; Gould et al., 2001; Kornack and Rakic, 2001a). Oral delivery may be preferred for studying adult neurogenesis, as invasive procedure may induce stress, a condition reported to affect the rate of neurogenesis (Gould et al., 1998). However, controlling the amount of BrdU absorbed per animal may represent a challenge (Zhao et al., 2003) (table. 1).

3.1.2. BrdU Availability in the CNS

The mechanism of transport of BrdU into the brain remains to be characterized. *In vitro* studies suggest that BrdU is transported by the same nucleoside transporter as thymidine (Lynch et al., 1977). Thymidine enters the brain primarily through facilitative low affinity, high-capacity carrier-mediated nucleoside transport system at the BBB, and may be eliminated through the cerebrospinal fluid, like thymidine (Thomas nee Williams and Segal, 1996; Thomas et al., 1997; Spector and Berlinger, 1982). It is estimated that BrdU is available for labeling in the adult brain for approximately a 2-h period, after systemic injection. After 2 h, the level of BrdU available for labeling in the brain drops sharply (Packard et al., 1973; Hayes and Nowakowski, 2000).

Table 1. Samples of studies on adult neurogenesis and their paradigms

Reference	Study	Paradigm (injection mode and BrdU dose)
Zhao et al., 2003	Neurogenesis SN	Single or repeated injections 100 mg/kg, i.p., drinking water, i.c.v.
Lie et al., 2002	Neurogenesis SN	Single or 10 days injections 50-100 mg/kg, i.p.
Cameron and McKay, 2001	Neurogenesis DG	Single injections 25, 50, 100, 300, 600 mg/kg, i.p.
van Praag et al. 1999	Running activity	12 days injections 50 mg/kg daily, i.p.
Bauer and Patterson, 2005	Abortive cell cycle re-entry	Continuous for 24 h 0.4 mg, i.c.v.
Kuan et al., 2004	Abortive cell cycle re-entry	Single injection 50 mg/kg, i.p.
Palmer et al., 2000	Neurogenesis DG	Single or repeated injections 50 mg/kg, i.p.
Kempermann et al. 1997	Environmental enrichment	12 days injections 50 mg/kg daily, i.p.
Gould et al. 1999a,b	Neurogenesis DG, primate study	Single or repeated injections 75 mg/kg, i.p.
Kornack and Rakic, 2001a	Neurogenesis OB, primate study	Single or repeated injections 50-75 mg/kg, i.v.
Kornack and Rakic, 1999	Neurogenesis DG, primate study	Single or repeated injections 50, mg/kg, i.v.

BrdU immunohistochemistry has been instrumental to confirm that neurogenesis occurs in the adult mammalian brain, including in human. BrdU may be delivered by intracerebroventricular (i.c.v.), intraperitoneal (i.p.) and intravenous (i.v.) injection, or orally, in drinking water, for studying neurogenesis. A single injection of thymidine analog labels nuclei of cells that are in S-phase, but not nuclei of proliferating cells that are in other phases of the cell cycle. Labeling the total proliferating population of cells in the brain at any on time require multiple injections every two hours, estimated time that BrdU is available for labeling in the brain, throughout the length of the cell cycle. Labeling the proliferating population of cells over an extended period of time requires multiple injections over several days or weeks. In this latter scenario, cells may be submitted to cumulative labeling, as they may divide a number of times during the labeling process, while cells dividing when BrdU is not present for their labeling will not be detectable. Hence, neither a single injection, nor repeated injections of thymidine analog provide a true estimate of the number of proliferating cells, but rather a relative number of proliferating cells in relation to the labeling paradigm. Most studies on adult neurogenesis in rodent and in primate use BrdU 50-100 mg/kg (body weight, i.p.). Data shows that 200 mg/kg (body weight, i.p.) is a saturating concentration of BrdU in rodent for studying adult neurogenesis.

3.1.3. BrdU a Tracer for Studying Neurogenesis

Protocols have been devised to estimate the number of proliferating cells, using thymine analogs. Thymidine analogs, like BrdU, integrate DNA of dividing cells during the S-phase of the cell cycle. Hayes and Nowakowski (2002) estimated the length of the cell cycle to 12-14 h. As the S-phase corresponds to an estimated third to half of the cell cycle length (Takahashi et al., 1995; Alexiades and Cepko, 1996) -over 4 h according to the estimation by Hayes and Nowakowski (2002)-, a single injection of thymidine analog labels nuclei of cells that are in S-phase, but not nuclei of proliferating cells that are in other phases of the cell cycle, during the period it is available for labeling in the adult brain, estimated at 2 h after systemic injection (Packard et al., 1973; Hayes and Nowakowski, 2000). Labeling the total proliferating population at any one time or proliferating cells over an extended period of time requires multiple injections of thymidine analog (Nowakowski et al., 1989). Labeling the total proliferating population of cells in the brain at any on time require multiple injections (i.p.) every 2 h throughout the length of the cell cycle. Labeling the proliferating population of cells over an extended period of time requires multiples injections over several days or weeks. In this latter scenario, on the one hand, cells may be submitted to cumulative labeling, as they may divide a number of times during the labeling process, while cells dividing when BrdU is not present for their labeling will not be detectable. On the other hand, multiple labeling may offer a more robust staining than single injection, as nucleoside analogs may be diluted and hence undetectable, as the cells divide further, once administration of BrdU is completed. In all, neither a single injection, nor repeated injections of thymidine analog provide a true estimate of the number of proliferating cells, but rather a relative number of proliferating cells in relation to the labeling paradigm, the length of the S-phase and of the cycle of the dividing cell population, the number of time the cells divide in presence of BrdU and their survival (Hayes and Nowakowski, 2002), as cell death is normally occurring process in the adult brain and particularly for newly generated cells in the neurogenic zones (Morshead and van der Kooy, 1992; Cameron and McKay, 2001).

3.1.4. BrdU Labeling Paradigm

Paradigms have been devised to determine the time of origin, migration, lineage and fate of neuronal cells in the developing and adult brain, using thymidine analogs like BrdU. This by varying the period between the pulse of thymidine analogs and the sacrifice of animals (Miller and Nowakowski, 1988). The determination of the time and site of origin of newly generated cells in the developing and adult CNS requires sacrifice the animals shortly, generally between 1-3 hours, after the administration of the thymidine analog; so that the migration of the labeled cells in S-phase is limited (Taupin et al., 2000). Cell migration is determined by sacrificing animals at various post-injection times, whereas the fate of the newly generated cells is determined by sacrificing the animals after their migration is completed (Cameron et al., 1993; Lois and Alvarez-Buylla, 1994; Palmer et al., 2000).

3.1.5. Fate of Newly Generated Neuronal Cells

The fate of newly generated neuronal cells is determined by multiple labeling and confocal microscopy. During development and adult neurogenesis, neuronal progenitor cells divide in discrete zones of proliferation or neurogenic areas, and migrate to their appropriate destination, while differentiating into their mature phenotypes and connecting their target cells. In the course of their maturation, newly generated neuronal cells express sequentially various markers that reflect their state of differentiation, from neural progenitor cells to terminally differentiated neurons (refer to section 2.2.). Numerous markers of neuronal cells for different stages of their maturation have been identified. They can be classified as markers of neural progenitor and stem cells, like nestin (Lendahl et al., 1990), Musashi 1 (Kaneko et al., 2000), sox-2 (Graham et al., 2003) and oct3/4 (Okuda et al., 2004), markers of immature neuronal cells or early post-mitotic neurons, like doublecortin (Francis et al., 1999; Gleeson et al., 1999), polysialylated-neural cell adhesion molecule (PSA-NCAM) (Bonfanti and Theodosis, 1994), Tuj-1 (Fanarraga et al., 1999), Hu (Marusich et al., 1994) and TUC-4 (Quinn et al., 1999), and markers of mature neurons, like calbindin-D28k (Sloviter, 1989), microtubule-associated protein-2 (Bernhardt and Matus, 1984) and NeuN (Mullen et al., 1992) (table 2). The identification of the phenotype of newly generated neuronal cells at the time of sacrifice is performed by multiple labeling, against BrdU and neuronal markers. Quantitative studies are performed by stereological analysis (West et al., 1991).

During development and adult neurogenesis, neuronal progenitor cells divide in discrete zones of proliferation or neurogenic areas, and migrate to their appropriate destination, while differentiating into their mature phenotypes and connecting their target cells. In the course of their maturation, newly generated neuronal cells express sequentially various markers that reflect their state of differentiation, from neural progenitor cells to terminally differentiated neurons. Numerous markers of neuronal cells at different stages of maturation have been identified. Markers of neural progenitor and stem cells, nestin, Musashi 1, sox-2, oct3/4, and immature neurons, doublecortin and Tuj-1, have been also detected in non-neuronal cells, like glial cells and tumor cells, or are re-expressed in preexisting neurons. Nonetheless, prospectively labeling newly generated cells and revealing their maturation and integration into the CNS network represent convincing evidence that neurogenesis occurs in the adult brain, using BrdU labeling.

Table 2. Markers for neural progenitor and stem cells, and immature neurons

Marker	Glial cells	Tumor cells
Neural progenitor and stem cells		
Nestin (Lendahl et al. 1990)	Frisen et al., 1995	Florenes et al., 1994
Musashi 1 (Kaneko et al., 2000)	Kanemura et al., 2001	
Sox-2 (Graham et al., 2003)	Komitova and Eriksson, 2004	
Oct 3/4 (Okuda et al., 2004)		Tai et al., 2005
Immature neurons		
Doublecortin (Francis et al., 1999; Gleeson et al., 1999)		Oltra et al., 2005
Tju-1 (Fanarraga et al., 1999)		Katsetos et al., 2001

Protocols, like single injection of BrdU and sacrificing the animals 2 h later provided valuable information on where cells divide and what their phenotypes were during S-phase (Taupin et al., 2000). Single injection of BrdU (50 mg/kg body weight, i.p.), four consecutive daily injections of BrdU or daily injections of BrdU for up to 10-12 days i.p. have proven successful for studying neurogenesis in the adult rodent DG (Kuhn et al., 1996; Cooper-Kuhn, 2002). Daily injections of BrdU for up to 10-12, sacrificing the animal 2 h after the last injection, days provided a protocol to observe cell proliferation and early migration (Palmer et al., 2000). Four consecutive daily injections of BrdU or daily injections of BrdU for up to 10-12 days, sacrificing the animals 4 weeks later provided a protocol to observe the phenotype of the newly generated cells after terminal differentiation in the adult DG, by allowing newly generated cells to undergo full neuronal maturation (Kuhn et al., 1996; Cooper-Kuhn, 2002).

In all, strategies have been devised to study cell proliferation, and adult neurogenesis using BrdU labeling. However, these paradigms are subject to numerous limitations and pitfalls. These issues will be discussed later in the manuscript (refer to section 4).

3.2. Cell Cycle Markers

The expression of markers of the cell cycle, like proliferating nuclear antigen (PCNA) and Ki-67 has been proposed to assay cell proliferation *in situ* (Kurki et al., 1986). PCNA, a cofactor of DNA polymerase, is expressed during the S-phase of cell cycle (Kurki et al., 1986). PCNA is also expressed in cell undergoing DNA repair, and some non-proliferating neurons (Hall et al., 1993; Ino and Chiba, 2000). Ki-67 is expressed in all phases of the cell cycle except the resting phase and a short period at the beginning of the phase G1 (Lopez et al., 1991; Endl and Gerdes, 2000; Zacchetti et al., 2003). Ki-67 has a very short half-life, is not detectable during DNA repair processes and is strongly downregulated/absent in quiescent cells (Takahashi and Caviness, 1993; Scholzen and Gerdes, 2000; Zacchetti et al.,

2003). In quiescent cells, Ki-67 is detected at sites linked to ribosomal RNA synthesis (Bullwinkel et al., 2006). Despite being detected in quiescent cells, Ki-67 offers a more reliable marker to identify cells that re-enter the cell cycle than PCNA (Tanapat et al, 1999; Kee et al., 2002). The quantification of Ki-67-positive cells has been shown to reflect cellular proliferation in a manner consistent with BrdU labeling in the adult DG (Eadie et al., 2005), supporting the use of markers of the cell cycle for studying adult neurogenesis. When using Ki-67 to quantify cell proliferation, significantly more cells will be labeled with Ki-67 than BrdU by immunohistochemistry, as the former is expressed during most phases of the cell cycle and the latter label only S-phase cells, thereby allowing a better estimation of the proliferative activity (Eadie et al., 2005). Other markers of the cell cycle, phosphorylated histone H3 and ribonucleotide reductase (RNR) have also proven to be valid candidates for studying cell proliferation (Engstrom et al., 1985; Hendzel et al., 1997; Zhu et al., 2003, 2005). Phosphorylated histone H3 is expressed during initial stages of chromatin condensation in late G2 interphase until anaphase (Hendzel et al., 1997). RNR is an enzyme for DNA synthesis composed of two subunits, M1 and M2. RNR expression peaks during S-phase and M1 is expressed by proliferating cells, but not quiescent ones (Engstrom et al., 1985; Mann et al., 1988).

However, the use of markers of the cell cycle for studying adult neurogenesis is limited by the temporal expression of cell cycle proteins that are only express during the phase(s) of the cell cycle, but not anymore when newly generated neuronal cells exit the cell cycle and begin their maturation process. It is also limited by the fact that cell cycle markers are most useful for arguing that a normally quiescent cell (G0) has re-entered the cell cycle and resumed DNA synthesis, but is not indicative that the cells completed the cell cycle.

3.3. Other Strategies

Other strategies have been devised and successfully applied for studying adult neurogenesis. In rodent, retroviral labeling and transgenic animal models provide alternative and complementary strategies for studying adult neurogenesis (van Praag et al., 2002; Yamaguchi et al., 2000). In human, most of these strategies are not applicable, or like BrdU labeling limited to cases when patients treated with BrdU as part of their cancer therapy, donated tissue samples for research investigations (Eriksson et al., 1998). Most studies in human rely therefore on the use of markers of the cell cycle, like PCNA and Ki-67, to study adult neurogenesis (Curtis et al., 2003; Sanai et al., 2004). Retrospective birth dating using [14C] may provide an additional mean to this aim (Spalding et al., 2005).

4. Assessment of BrdU Labeling for Studying Cell Proliferation and Adult Neurogenesis

BrdU labeling is currently the most used method for studying adult neurogenesis. There are however pitfalls and limitations to the use of BrdU immunohistochemistry for studying cell proliferation, and neurogenesis.

4.1. BrdU a Toxic Substance

BrdU is a toxic substance. In rodent, when injected into pregnant animals, BrdU may cause exencephaly, cleft palate, limb abnormalities, and can lead to teratogenic malformations and behavioral changes in the progenies (Bannigan and Langman, 1979; Bannigan, 1985; Bannigan et al., 1990; Nagao et al., 1998; Kuwagata and Nagao, 1998; Kolb et al., 1999). When administered postnatally, BrdU may cause lung changes in mature rodent, and alter the development of the cerebellum (Yu, 1976; Nagai et al., 1993). In the CNS, high doses of BrdU 60-600 mg/kg (body weight, i.p.) trigger neuronal cell death during embryonic and neonatal development (Bannigan, 1985; Nagao et al., 1998). Although lower doses of BrdU 50 mg/kg (body weight, i.p, single injection) have no apparent toxic effect on development of the cortical ventricular zone (Miller and Nowakowski, 1988; Takahashi et al., 1995), administration of multiple doses of BrdU 12-20 mg/kg in pregnant rats affects the litter size, body weight and mortality of the offsprings. It also reduces the size of the cerebellum, and produces defects in proliferation, migration and patterning of the cerebellum in the adult progenies (Sekerkova et al., 2004).

In adult, most investigators use BrdU 50-100 mg/kg (body weight, i.p.) to study neurogenesis in rodents and non-human primates (Corotto et al., 1993; Seki and Arai, 1993; Kuhn et al., 1996; Gould et al., 2001; Kornack and Rakic, 2001a; van Praag et al., 1999). BrdU 50 mg/kg (body weight, i.p.) for up to 12 days, as well as doses as high as 300 mg/kg have no physiological side effects, like weight loss or behavioral changes in adult rats, and no apparent toxic effect on dividing cells in the DG (Cameron and McKay, 2001; Cooper-Kuhn and Kuhn, 2002). Administration of BrdU would have less toxicity for the adult CNS than embryonic and neo-natal CNS, as the BBB is more developed in adult. The BBB develops around the time of parturition, post-natal day 10 in rat (Ribatti et al., 2006). Nonetheless, BrdU is a toxic substance. The integration of halogenated thymine analogs into the DNA alters its stability, increasing the risk of sister-chromatid exchanges, mutations, DNA double-strand breaks, and lengthens the cell cycle of cells that incorporate it (Bannigan and Langman, 1979; Saffhill and Ockey, 1985; Morris, 1991, 1992). Therefore, although no apparent toxic effect of BrdU have been reported when studying adult neurogenesis, even low doses of BrdU are likely to have toxic effects on newly generated neuronal cells in the adult brain, particularly when BrdU is administered orally or multiple doses are administered (Sekerkova et al., 2004).

4.2. Saturating Doses of BrdU

Adult neurogenesis is affected by various conditions, like environment, trophic factors and drug treatments, various physiopathological conditions, as aging, hormones, diseases, disorders and injuries, and genetic background (van Praag et al., 2000; Taupin, 2005). Some of these conditions, like anesthetics and glucocorticoid treatments (Endo et al., 1997), exercise (Sharma et al., 1991; Ide and Secher, 2000; Delp et al., 2001; Swain et al., 2003), kainic-acid treatments and seizures (Pont et al., 1995; Carpentier et al., 1990), stress (Tillfors et al., 2001), stroke, traumatic brain injury, disrupt the BBB permeability and/or alter the blood flow to the brain. An increase in BrdU labeling in the brain could originate from an increase in BrdU uptake rather than an increase in cell proliferation and neurogenesis (Gould and Gross, 2002; Eadie et al. 2005). An increase in BrdU uptake may also underlies differences in BrdU labeling in the brain when comparing different genetic background, species, and adult and developmental stages, as different species have differences in the permeability of the BBB to many compounds and the BBB develops around the time of parturition (Thomas and Segal, 1997; Ballabh et al., 2004; Ribatti et al., 2006). Determining the saturating concentration of BrdU needed to label most S-phase cells in the brain may allow comparison between groups.

On the one hand, Cameron and McKay (2001) reported that single doses of BrdU 100, 50 and 25 mg/kg (body weight, i.p.) labels 60%, 45% and 8% of S-phase cells in adult rat DG, respectively. Doses greater than or equal to 300 mg/kg are needed to label most S-phase cells, as the number of BrdU-labeled cells appeared to plateau (Cameron and McKay, 2001). On the other hand, according to Burns and Kuan (2005) doses of 50-100 mg/kg BrdU (body weight, i.p.) are sufficient to label the vast majority of S-phase proliferative cells in mice (Burns and Kuan, 2005). A recent study by Eadie et al. (2005) showed that single doses of BrdU 200 mg/kg (body weight, i.p.) and above are saturating in rat, confirming the data of Cameron and McKay (2001). BrdU 200 mg/kg also labels significantly more cells in animals that exercise (Eadie et al., 2005). In all, these data shows that 200 mg/kg (body weight, .p.) is a saturating concentration of BrdU to use in rodent for studying adult neurogenesis when comparing groups. As BrdU is a toxic substance, lower doses may be use for analytical studies of neurogenesis (Sekerkova et al., 2004).

4.3. DNA Repair

DNA repair is a normally occurring process in the life of a cell and is carried out by cellular enzymes, which secure genomic stability (Memo, 1999). Because DNA repair involves DNA synthesis and because BrdU is not a marker for cell proliferation, but a marker of DNA synthesis, there is a concern that BrdU immunohistochemistry may not only detect dividing cells in the brain, but also cells undergoing DNA repair. The level of DNA repair in the normal brain, as measured by [3H]-thymidine autoradiography, is remarkably constant throughout all neurons of the hippocampus, as well as in other neuronal populations of the brain (Schmitz et al., 1999). In contrast, neurogenesis occurs primarily in two areas of the adult brain, the DG and the SVZ (Taupin and Gage, 2002). This suggests that standard protocols to study neurogenesis in the adult brain by BrdU immunohistochemistry are not sensitive enough to label cell undergoing DNA repair. This is likely because DNA repair occurs normally *in vivo* through a mechanism that replaces only 1–2 nucleotides at each site, as opposed to cell division where the entire genome is replicated.

DNA repair often occurs in neuronal cells after irradiation or exposure to mutation-inducing chemicals, as an attempt to rescue post-mitotic neurons (Gobbel et al., 1998; Lowndes and Murguia, 2000; Korr et al., 1989). A high percentage of these cells will undergo apoptotic cell death. DNA damaged-induced by ionizing radiation are dose-dependent, and repaired through a pathway that replaces approximately 100 nucleotides at each site (Shinohara et al., 1997; Korr et al., 1989; Li et al., 2001a,b). The detection of DNA repair by BrdU immunohistochemistry has been reported *in vitro* on irradiated fibroblasts. It requires a 24-48 h exposure to BrdU (Beisker and Hittelman, 1988; Selden et al., 1993, 1994). Using standard protocols to study adult neurogenesis by BrdU labeling, Palmer et al. (2000) did not detect radiation-induced DNA repair on fibroblasts in culture. Further, *in vivo*, fractionated brain irradiation is associated with a reduction in the number of newly generated neuronal cells, and a decline in BrdU labeling in the neurogenic areas, the ventricle wall and the DG, in rodent (Parent et al., 1999; Santarelli et al., 2003). The reduction in the number of newly generated neuronal cells and decline in BrdU labeling are dose-dependent with increasing doses of radiation (Tada et al., 1999, 2000; Mizumatsu et al., 2003). In all, these data show that BrdU staining may not reflect DNA repair in most experimental conditions used to study adult neurogenesis.

4.4. Gene Regulation

In vitro, BrdU prevents differentiation of post-natal and embryonic cells without affecting cell division or cell viability (Wilt and Anderson, 1972; Younkin and Silberberg, 1973, 1976; Fukushima and Barka, 1976; Nakashima et al., 1984), and increases cell proliferation, myeloid differentiation and the ability of adult bone marrow-derived mesenchymal stem cells to differentiate into neural and retinal cells (Keoffler et al., 1983; Qu et al., 2004). These data show that BrdU can induce or enhance cell proliferation, and alter the phenotype of cells. These activities mediated at the transcriptional and/or translational levels have profound consequences on cell proliferation and fate determination when using BrdU labeling for studying adult neurogenesis, and the phenotype of grafted cells pre-labeled with BrdU. Report by Qu et al. (2004) suggests that BrdU may affect the mitotic activity and phenotype of those cells. Particularly, adult stem cells are multipotent; they give rise to phenotype restricted to the tissues from which they are derived. Over the past years, numerous studies have reported that adult stem cells may have a broader potential, they may give rise to lineages other than their tissues of origin (Anderson et al., 201; D'Amour and Gage, 2002; Mezey, 2004). Some of these studies have used BrdU-prelabeled cells to show such broader potential after transplantation (Kopen et al., 1999; Li et al., 2001a,b). Report by Qu et al. (2004) suggests that BrdU may affect the phenotype of the grafted cells, and that the broader potential observed may not be due to the intrinsic properties of the grafted cells to differentiate in a new environment, but it may be enhanced or induced by BrdU. Therefore, investigators must be cautious when analyzing and interpreting studies in vivo and ex vivo, on cell division and differentiation, when using BrdU labeling.

4.5. Abortive Cell Cycle Re-entry

Cell death is a normally occurring process in the adult brain especially in the neurogenic zones, as a significant proportion of newly generated cells in the adult SVZ and SGZ are believed to undergo apoptosis rather than achieving maturity (Cameron and McKay, 2001;

Morshead and van der Kooy, 1992). Damaged or degenerating terminally differentiated post-mitotic neurons in the adult brain can re-enter the cell cycle, activate cell cycle-associated proteins, cyclins and ubiquitins, and initiate abortive DNA synthesis without cell division, before undergoing cell death (Chen et al., 2000; Katchanov et al., 2001; Liu and Greene, 2001; Yang et al., 2001; Herrup and Arendt, 2002). Although cell cycle re-entry is not a common mechanism of neurodegeneration, since most dying neurons do not pass the G1/S-phase checkpoint to resume DNA synthesis (Snider et al., 1999), it could lead to false positive when using BrdU for studying adult neurogenesis, as BrdU incorporates DNA of dividing cells during the S-phase of the cell cycle (Kuan et al., 2004). The significant proportion of newly generated neuronal progenitor cells undergoing cell death rather than achieving maturity in the intact adult brain, and after injury -80% of the new striatal neuronal cells that are generated from the SVZ after stroke in rat die within the first weeks after the insult- (Cameron and McKay, 2001; Morshead and van der Kooy, 1992; Arvidsson et al., 2001; Sun et al., 2004) suggests that it may provide a window of opportunity when cell undergoing programmed cell death or abortive cell cycle re-entry, as prelude to apoptosis, can be rescued and directed to participate in the regeneration of the damaged tissue. Factors preventing cell death, like caspases (Pompeiano et al., 2000; Ekdahl et al., 2001), would be potentially beneficial for cellular therapy, alone or in combination with the administration of trophic factors. However, though caspases have shown to promote survival of newly generated neuronal cells in the CNS, there is no proof of the long-term survival and functionality of these cells (Pompeiano et al., 2000; Ekdahl et al., 2001).

Two studies have recently investigated whether dying post-mitotic neurons could undergo abortive cell cycle re-entry and incorporate thymidine analogs. Kuan et al. (2004) reported that the combination of hypoxia and ischemia –but not independently- in mice induces terminally differentiated neurons to reenter the cell cycle, and incorporate BrdU. In their study, Kuan et al. (2004) detected BrdU, NeuN, a marker of mature neurons, and terminal deoxynucleotidyl transferase-mediated biotinylated dUTP nick end-labeling (TUNEL)-positive cells for 5 days after BrdU injection, that eventually disappeared after 28 days, suggesting that the combination hypoxia-ischemia triggers neurons to reenter the cell cycle and resume apoptosis-associated DNA synthesis in brain. Bauer and Patterson (2005) did not detect BrdU in TUNEL-positive cells in three mouse models displaying concomitant stimulation of apoptosis and neurogenic proliferation, olfactory bulbectomy, brain irradiation and kainic acid-induced seizure, suggesting that experimental models that induce neuronal cell death may not be efficient in inducing cell cycle reactivation of post-mitotic neurons. It is hypothesized that the combination of hypoxia and ischemia could synergistically activate specific components of the DNA damage response that would trigger aberrant cell cycle reactivation in post-mitotic neurons (Kuan et al. 2004). Therefore, though dying post-mitotic neurons may rarely progress through S-phase or incorporate BrdU, it remains critical that appropriate controls be performed when studying adult neurogenesis, particularly investigating cell death in conjunction with BrdU labelling. Cell death can be studied *in situ* by the expression of proteases, like caspases and by TUNEL assay (Namura et al., 1998; Gavrieli et al., 1992). Caspase or TUNEL labeling in conjunction with BrdU labelling is therefore most useful for arguing that a normally quiescent cell (G0) has re-entered the cell cycle and resumed DNA synthesis, through cell proliferation or abortive cell cycle re-entry.

However, the use of cell death marker may only be useful in the few hours-days following the injection of BrdU, as cell death is a transient process leading to the elimination of the cells (Kuan et al., 2004).

4.6. Gene Duplication

The adult brain contains a substantial fraction of cells that are aneuploids (5-7% in mice) (Rehen et al., 2001, 2005). The origin of these cells remains to be determined. Aneuploid cells may originate from cell cycle re-entry and DNA duplication. Alternatively, they may originate from cell fusion (Alvarez-Dolado et al., 2003). Vincent et al. (1996) reported by immunohistochemistry the expression of proteins of the cell cycle, like cyclin B a marker of the phase G2, in neurons in regions of human brain from patients with AD. These cell cycle proteins were detected in regions in which degeneration occurs, like the hippocampus (Vincent et al., 1996). Yang et al. (2001) using fluorescence *in situ* hybridization (FISH) showed that some at-risk neurons in the AD brain are aneuploids. The characterization of cyclin B and aneuploid cells suggests that cells entered the cell cycle and underwent DNA replication, but did not complete the cell cycle: the cells completed phases G1, S and G2, but not M of the cell cycle. Four to 10% of the neurons in regions of degeneration were reported expressing cell cycle proteins or tetraploids (Busser et al., 1998; Yang et al., 2001). It is proposed that the genetic imbalance in tetraploid cells signifies that they are fated to die (Yang et al., 2001). Their relatively high percentage at any one time suggests that they will undergo a slow death process, unlike apoptosis; these cells may live in this state for months, possibly up to 1 year (Busser et al., 1998; Yang et al., 2001). Cell cycle re-entry in region of neurodegeneration has been reported in experimental models of lesion-induced cell death, by experimentally expressing cell cycle proteins in a mature neurons, leading to cell death rather than cell division, as well as in other pathological conditions that involve loss of neuronal function, like strokes, amyotrophic lateral sclerosis, some forms of encephalitis and mild cognitive impairment (Herrup et al., 2004). It is hypothesized that in the diseased and injured brain, the disregulation and/or re-expression of proteins controlling the cell cycle would cause DNA synthesis. Therefore, beside abortive cell cycle re-entry, cell cycle re-entry and DNA duplication also precedes neuronal death in degenerating regions of the CNS. The mechanisms underlying cell cycle re-entry and its blockage in phase G2 are mostly unknown. β-Amyloid, reactive oxygen species (ROS) and oxidative stress are candidates to induce cell cycle re-entry and neuronal death (Haughey et al., 2002; Klein and Ackerman, 2003; Langley and Ratan, 2004). In support of this contention, the Harlequin (Hq) mutation, an X-linked mutation in mouse, is characterized by neuronal cell death in cerebellum, in which neurons reenter the cell cycle, as evidence of PCNA immunostaining and BrdU-labeling (Klein et al., 2002). Hq mutation is caused by proviral insertion in the first intron of the apoptosis-inducing factor (Aif) gene, and results in downregulation of the expression of AIF and increase in oxidative damage. This suggests that AIF regulates the amount of ROS, and ROS might initiate cell cycle re-entry in postmitotic neurons.

The evidence that cell cycle re-entry and DNA duplication also precedes neuronal death in degenerating regions of the CNS has profound implications when studying neurogenesis in animal models of CNS diseases and injuries, and in the diseased and injured human brain. Neurons in degenerating regions of the CNS can activate cell cycle proteins, cyclins and ubiquitins, and initiate DNA synthesis without cell division, leading to polyploidy that can persist for many months before the death of the cells (Chen et al., 2000; Katchanov et al., 2001; Liu and Greene, 2001; Yang et al., 2001; Herrup and Arendt, 2002). As BrdU incorporates DNA of dividing cells during the S-phase of the cell cycle, BrdU labeling will not allow distinguishing between cell proliferation, cell cycle re-entry and DNA duplication without cell division, neither will immunohistochemistry for markers of the cell cycle. Several studies have reported the generation of new neuronal cells in experimental models of injuries and diseases, and in the human diseased brain. Magavi et al. (2000) used targeted apoptosis of cortical pyramidal neurons by photolytic lesion in mice, and immunostaining for BrdU, doublecortin, Hu, and NeuN. The authors found a small number of new neurons extending processes to the original site of degeneration in the thalamus. Jin et al. (2004a) reported an increase in the expression of markers for immature neuronal cells, doublecortin, PSA-NCAM, in the SGZ and the granular layer of the DG, as well as in the CA1 region of the hippocampus, from autopsies of AD brain patients. Studies from animal models of AD using BrdU labeling revealed that neurogenesis is positively regulated in the DG of transgenic mice that express an amyloid precursor protein mutation found in some familial forms of AD (Jin et al., 2004b). In these studies, the existence of aneuploid cells has not been investigated, and may account for some of the observations reported by the authors as newly generated neuronal cells. Therefore, future investigations will aim at confirming that neurogenesis is induced in the diseased and injured brain. This will require verifying that the incorporation of BrdU does not represent aneuploidy. Immunohistochemistry for markers of the cell cycle is indicative that a normally quiescent cell (G0) has re-entered the cell cycle and resumed DNA synthesis, but not that the cells completed the cell cycle. Therefore, the best control is to quantify the number of copies of chromosomes, by FISH (Blaschke et al., 1996; Rehen et al., 2001, 2005). This will allow the distinction of cell proliferation versus cell cycle re-entry and DNA duplication without cell division, leading to polyploidy.

4.7. BrdU immunohistochemistry and Multiple Labeling

BrdU immunohistochemistry has several limitations for studying cell proliferation, and adult neurogenesis. Protocols to detect BrdU by immunohistochemistry are devised using monoclonal antibody directed against single stranded DNA containing BrdU (Gratzner, 1982). The use of primary antibodies directed against BrdU on single-stranded DNA requires denaturing DNA. Standard procedure for denaturing the DNA involves partial hydrolysis of the tissue with HCl treatment (Gratzner. 1982; Moran et al., 1985; del Rio and Soriano, 1989). Such treatment can affect the cell morphology and antigenicity recognition in multiple labeling studies, thereby limiting the morphological and phenotypical identification of newly generated cells and their phenotype identification. Studies describing the main cell types relevant to adult neurogenesis have been identified using [3H]-thymidine autoradiography and in transgenic mice expressing green fluorescent protein under the nestin promoter (Filippov et al., 2003). Particularly, using [3H]-thymidine autoradiography, Doetsch et al.

(1997) and; Garcia-Verdugo et al. (1998) reported the existence of three cell types, type A, B and C cells, relevant to adult neurogenesis in the rodent SVZ. Protocols to denature DNA without the deproteinizing effect of HCl treatment have been proposed using DNA fragmentation with DNase I, and enable a broader range of antigen detection in multiple labeling studies (Sekerkova et al., 2004). Denaturing DNA also prevents its labeling with fluorescent probes, like propidium iodide (PI) and DAPI (4',6-Diamidine-2'-phenylindole). PI and DAPI bind double-stranded DNA. These markers are useful for visualizing the nucleus, and the state of the cell cycle, particularly metaphases. Although BrdU labeling confer significant advantages over [3H]-thymidine labeling; it allows faster studies without handling radiolabeled material and the detection of labeled cells throughout the relatively thick tissue sections required for stereological studies of the brain (West et al., 1991), BrdU immunohistochemistry, in contrast to [3H]-thymidine autoradiography, is not stoichiometric. The intensity or extent of BrdU-labeling is highly dependent on the methods used for detection, and does not necessarily reflect the magnitude of DNA replication (Nowakowski and Hayes, 2000).

Immunohistochemistry and confocal microscopy protocols have been developed to co-label BrdU and other markers of the neuronal lineages. Prospectively labeling newly generated cells, and reveling their maturation and integration into the CNS network, using multiple labeling and BrdU, has proven valuable for characterizing adult neurogenesis. There are mainly two caveats in this approach. The first one relates to the specificity of the markers used for identifying neural progenitor and stem cells, and immature neurons. Nestin, musashi 1, sox-2, oct3/4, doublecortin, PSA-NCAM and Tuj-1 are markers of neural precursor cells. They have been also detected in non-neuronal cells, as in glial cells (Frisen et al., 1995; Komitova and Eriksson, 2004) and tumor cells (Florenes et al., 1994; Kanemura et al., 2001; Ignatova et al., 2002; Katsetos et al., 2001, 2003; Oltra et al., 2005; Tai et al., 2005), and are re-expressed in preexisting neurons (Charles et al., 2002) (table 2). Therefore, the phenotypic identification of a cell type based solely of immunohistochemistry may not reveal its exact identity. Electrophysiological studies may permit to functionally define the specified cell types. Nonetheless, time course studies revealing the sequential and progressive maturation of BrdU labeled-newly generated neuronal cells from neural progenitor cells to terminally differentiated neurons provide the strongest argument using BrdU labeling that neurogenesis occurs in adult brain. The second caveat is associated with identifying unequivocally the phenotype of the newly generated cells by confocal microscopy. Scanning throughout the thickness of the marked cells and recreating 3-D pictures of the cells is required to show unequivocal double labeling with BrdU and cell type-specific markers when using confocal microscopy.

4.8. BrdU Transfer

BrdU labeling is used as a marker in transplantation studies. Cells, including neural progenitor and stem cells, are labeled in culture by uptake of BrdU, prior to grafting. Upon sacrificing the animals, grafted cells are then detected by immunohistochemistry for BrdU to reveal the integration and phenotype of the grafted cells into the host (Gage et al., 1995;

Suhonen et al., 1996; Kopen et al., 1999; Shihabudin et al., 2000). A recent report shows that BrdU can be transferred from the graft to the genome of the host's adult neuronal cells (Burns et al., 2006). In this study, progenitor cells isolated from the adult bone marrow and fibroblasts were labeled in culture by BrdU or other halogenated thymidine analogs, like chlorodeoxyuridine. To control for the possibility that thymidine analog may be incorporated into dividing host cells from transplanted dying cell, grafted cells were killed by freeze-thawed prior transplantation. After transplantation in the SVZ, large number of neurons and glia were found labeled with thymidine analog 3 to 12 weeks after transplantation, whether bone marrow stem cells, fibroblasts or dead cells were transplanted. This shows that thymidine analog may be transferred from the graft to endogenous dividing cells. Since death of grafted cells is a well documented phenomenon during transplantation, the source of thymidine analog may likely originate from those cells in the graft.

Adult stem cells are multipotent; they give rise to phenotype restricted to the tissues from which they are derived. Numerous studies have reported that adult stem cells may have a broader potential, they may give rise to lineages other than their tissues of origin (Anderson et al., 2001; D'Amour and Gage, 2002; Mezey, 2004). Some of these studies have used thymidine analog-prelabeled cells to show such broader potential after transplantation. Particularly, marrow stromal stem cells or mesenchymal stem cells grafted into the brain have been shown to give rise to neuronal phenotypes, as evidence by the detection of BrdU-positive cells expressing neuronal markers (Kopen et al., 1999; Li et al., 2001a,b). Reports by Burns et al. (2006) suggests that some of these data may be due to the transfer of thymidine analog from the graft that is then being incorporated by dividing neighboring cells, leading to false positive. Data from Burns et al. (2006) further suggests that transfer of thymidine analog may also occurs when using thymidine analog to label endogenous progenitor and stem cells in the CNS. The most common fate of newly generated cells in the normal adult brain and after injuries is apoptosis (Morshead and van der Kooy, 1992; Cameron and McKay, 2001; Arvidsson et al., 2001; Sun et al., 2004). Endogenous labeled newly generated cells undergoing cell death would release thymidine that would be picked up by newly generated cells. Though this does not question endogenous proliferation, it may affect the temporal resolution of birth of the dividing population of cells (Burns et al., 2006). These data limit the use of BrdU and thymidine analogs as markers for grafted cells, and also for the study of endogenous newly generated cells. They mandate the need for appropriate controls. Other labeling strategies may be applied, like the use of genetically engineered cells expressing reported genes, for example the gene coding for green fluorescent protein, or Y/X chromosome discrimination in grafting experiments and retroviruses to study endogenous dividing cells.

5. Concluding Remarks

The advent of BrdU immunohistochemistry has significantly contributed to the confirmation that neurogenesis occurs in the adult mammalian brain. However, the use of BrdU labeling is not without pitfalls and limitations. BrdU is a toxic and mutagenic substance that has side effects. BrdU lengthens the cell cycle, and has mitogenic, transcriptional and translational effects. BrdU and thymidine analogs are markers of DNA synthesis, and not markers of the S-phase of the cell cycle or cell division. BrdU incorporates DNA during its repair, abortive cell cycle re-entry as prelude to apoptosis, and cell cycle re-entry and DNA duplication without cell division in the CNS. BrdU can be transferred from dying cells to neighboring dividing cells. BrdU is currently the most used technique to study adult neurogenesis. Although time course studies revealing the sequential and progressive maturation of BrdU labeled-newly generated neuronal cells from neural progenitor cells to terminally differentiated neurons provide the strongest argument using BrdU labeling that neurogenesis occurs in adult brain, the pitfalls and limitations discussed in this manuscript caution against the use of BrdU as the sole evidence to report the generation of new neuronal cells in the adult brain. Data must be carefully interpreted. Appropriate controls must be performed to ensure the incorporation of BrdU reflects the generation of new neurons in the adult mammalian brain, and alternative strategies must be considered.

Acknowledgments

Reproduced from: Taupin P. BrdU Immunohistochemistry for Studying Adult Neurogenesis: paradigms, pitfalls, limitations, and validation. Brain Research Reviews (2007) 53(1): 198-214. Copyright (2007), with permission from Elsevier.

References

Alexiades MR, Cepko C. (1996) Quantitative analysis of proliferation and cell cycle length during development of the rat retina. *Dev. Dyn.* 205, 293-307.

Altman J, Das GD. (1965) Autoradiographic and histological evidence of postnatal hippocampal neurogenesis in rats. *J. Comp. Neurol.* 124, 319-36.

Altman J. (1969) Autoradiographic and histological studies of postnatal neurogenesis. IV. Cell proliferation and migration in the anterior forebrain, with special reference to persisting neurogenesis in the olfactory bulb. *J. Comp. Neurol.* 137, 433-58.

Altman J, Bayer SA. (1990) Mosaic organization of the hippocampal neuroepithelium and the multiple germinal sources of dentate granule cells. *J. Comp. Neurol.* 301, 325-42.

Alvarez-Buylla A, Garcia-Verdugo JM, Tramontin AD. (2001) A unified hypothesis on the lineage of neural stem cells. *Nat. Rev. Neurosci.* 2, 287-93.

Alvarez-Dolado M, Pardal R, Garcia-Verdugo JM, Fike JR, Lee HO, Pfeffer K, Lois C, Morrison SJ, Alvarez-Buylla A. (2003) Fusion of bone-marrow-derived cells with Purkinje neurons, cardiomyocytes and hepatocytes. *Nature* 425, 968-73.

Anderson DJ, Gage FH, Weissman IL. (2001) Can stem cells cross lineage boundaries? *Nat. Med.* 7, 393-5.

Angevine JB Jr. (1965) Time of neuron origin in the hippocampal region. An autoradiographic study in the mouse. *Exp. Neurol.* 11, 1-70.

Arvidsson A, Kokaia Z, Lindvall O. (2001) N-methyl-D-aspartate receptor-mediated increase of neurogenesis in adult rat dentate gyrus following stroke. *Eur. J. Neurosci.* 14, 10-8.

Arvidsson A, Collin T, Kirik D, Kokaia Z, Lindvall O. (2002) Neuronal replacement from endogenous precursors in the adult brain after stroke. *Nat. Med.* 8, 963-70.

Ballabh P, Braun A, Nedergaard M. (2004) The blood-brain barrier: an overview: structure, regulation, and clinical implications. *Neurobiol. Dis.* 16, 1-13.

Bannigan JG. (1985) The effects of 5-bromodeoxyuridine on fusion of the cranial neural folds in the mouse embryo. *Teratology* 32, 229-39.

Bannigan J, Langman J. (1979) The cellular effect of 5-bromodeoxyuridine on the mammalian embryo. *J. Embryol. Exp. Morphol.* 50, 123-35.

Bannigan JG, Cottel DC, Morris, A. (1990) Study of the mechanisms of BUdR-induced cleft palate in the mouse. *Teratology* 42, 79-89.

Bauer S, Patterson PH. (2005) The cell cycle-apoptosis connection revisited in the adult brain. *J. Cell Biol.* 171, 641-50.

Bayer SA, Yackel JW, Puri PS. (1982) Neurons in the rat dentate gyrus granular layer substantially increase during juvenile and adult life. *Science* 216, 890-2.

Bedard A, Cossette M, Levesque M, Parent A. (2002) Proliferating cells can differentiate into neurons in the striatum of normal adult monkey. *Neurosci. Lett.* 328, 213-6.

Bedard A, Parent A. (2004) Evidence of newly generated neurons in the human olfactory bulb. *Brain Res. Dev. Brain Res.* 151, 159-68.

Bedard A, Gravel C, Parent A. (2006) Chemical characterization of newly generated neurons in the striatum of adult primates. *Exp. Brain Res.* 170, 501-12.

Begg AC, Hofland I, Van Der Pavert I, Van Der Schueren B, Haustermans K. (2000) Use of thymidine analogues to indicate vascular perfusion in tumours. *Br. J. Cancer* 83, 899-905.

Beisker W, Hittelman WN. (1988) Measurement of the kinetics of DNA repair synthesis after uv irradiation using immunochemical staining of incorporated 5-bromo-2'-deoxyuridine and flow cytometry. *Exp. Cell Res.* 174, 156-67.

Belluzzi O, Benedusi M, Ackman J, LoTurco JJ. (2003) Electrophysiological differentiation of new neurons in the olfactory bulb. *J. Neurosci.* 23, 10411-8.

Bernhardt R, Matus A. (1984) Light and electron microscopic studies of the distribution of microtubule-associated protein 2 in rat brain: a difference between dendritic and axonal cytoskeletons. *J. Comp. Neurol.* 226, 203-21.

Bernier PJ, Bedard A, Vinet J, Levesque M, Parent A. (2002) Newly generated neurons in the amygdala and adjoining cortex of adult primates. *Proc. Natl. Acad. Sci. USA* 99, 11464-9.

Blaschke AJ, Staley K, Chun J. (1996) Widespread programmed cell death in proliferative and postmitotic regions of the fetal cerebral cortex. *Development* 122, 1165-74.

Bonfanti L, Theodosis DT. (1994) Expression of polysialylated neural cell adhesion molecule by proliferating cells in the subependymal layer of the adult rat, in its rostral extension and in the olfactory bulb. *Neurosci.* 62, 291-305.

Brezun JM, Daszuta A. (1999) Depletion in serotonin decreases neurogenesis in the dentate gyrus and the subventricular zone of adult rats. *Neurosci.* 89, 999-1002.

Brown JP, Couillard-Despres S, Cooper-Kuhn CM, Winkler J, Aigner L, Kuhn HG. (2003) Transient expression of doublecortin during adult neurogenesis. *J. Comp. Neurol.* 467, 1-10.

Bullwinkel J, Baron-Luhr B, Ludemann A, Wohlenberg C, Gerdes J, Scholzen T. (2006) Ki-67 protein is associated with ribosomal RNA transcription in quiescent and proliferating cells. *J. Cell Physiol*. 206, 624-35

Burns KA, Kuan CY. (2005) Low doses of bromo- and iododeoxyuridine produce near-saturation labeling of adult proliferative populations in the dentate gyrus. *Eur. J. Neurosci.* 21, 803-7.

Burns TC, Ortiz-Gonzalez XR, Gutierrez-Perez M, Keene CD, Sharda R, Demorest ZL, Jiang Y, Nelson-Holte M, Soriano M, Nakagawa Y, Luquin MR, Garcia-VerdugoJM, Prosper F, Low WC, Verfaillie CM. (2006) Thymidine Analogs areTransferred from Pre-Labeled Donor to Host Cells in the Central Nervous SystemAfter Transplantation: A Word of Caution. *Stem Cells* 24, 1121-7.

Busser J, Geldmacher DS, Herrup K. (1998) Ectopic cell cycle proteins predict the sites of neuronal cell death in Alzheimer's disease brain. *J. Neurosci*. 18, 2801-7.

Cameron HA, Woolley CS, McEwen BS, Gould E. (1993) Differentiation of newly born neurons and glia in the dentate gyrus of the adult rat. *Neurosci.* 56, 337-44.

Cameron HA, McKay RD. (2001) Adult neurogenesis produces a large pool of new granule cells in the dentate gyrus. *J. Comp. Neurol*. 435, 406-17.

Carlen M, Cassidy RM, Brismar H, Smith GA, Enquist LW, Frisen J. (2002) Functional integration of adult-born neurons. *Curr. Biol.* 12, 606-8.

Carpentier P, Delamanche IS, Le Bert M, Blanchet G, Bouchaud C. (1990) Seizure-related opening of the blood-brain barrier induced by soman: possible correlation with the acute neuropathology observed in poisoned rats. *Neurotoxicology* 11, 493-508.

Charles P, Reynolds R, Seilhean D, Rougon G, Aigrot MS, Niezgoda A, Zalc B, Lubetzki C. (2002) Re-expression of PSA-NCAM by demyelinated axons: an inhibitor of remyelination in multiple sclerosis? *Brain* 125, 1972-9.

Chen Y, McPhie DL, Hirschberg J, Neve .L. (2000) The amyloid precursor protein-binding protein APP-BP1 drives the cell cycle through the S-M checkpoint and causes apoptosis in neurons. *J. Biol. Chem*. 275, 8929-35.

Chen J, Magavi SS, Macklis JD. (2004) Neurogenesis of corticospinal motor neurons extending spinal projections in adult mice. *Proc. Natl. Acad. Sci. USA* 101, 16357-62.

Cooper-Kuhn CM, Kuhn HG. (2002) Is it all DNA repair? Methodological considerations for detecting neurogenesis in the adult brain. *Brain Res. Dev. Brain Res*. 134, 13-21.

Corotto FS, Henegar JA, Maruniak JA. (1993) Neurogenesis persists in the subependymal layer of the adult mouse brain. *Neurosci. Lett*. 149, 111-4.

Craig CG, Tropepe V, Morshead CM, Reynolds BA, Weiss S, van der Kooy D. (1996) In vivo growth factor expansion of endogenous subependymal neural precursor cell populations in the adult mouse brain. *J. Neurosci.* 16, 2649-58.

Crespo D, Stanfield BB, Cowan WM. (1986) Evidence that late-generated granule cells do not simply replace earlier formed neurons in the rat dentate gyrus. *Exp. Brain Res.* 62, 541-8.

Curtis MA, Penney EB, Pearson AG, van Roon-Mom WM, Butterworth NJ, Dragunow M, Connor B, Faull RL. (2003) Increased cell proliferation and neurogenesis in the adult human Huntington's disease brain. *Proc. Natl. Acad. Sci. USA* 100, 9023-7.

D'Amour KA, Gage FH. (2002) Are somatic stem cells pluripotent or lineage-restricted? *Nat. Med.* 8, 213-4.

Danilov AI, Covacu R, Moe MC, Langmoen IA, Johansson CB, Olsson T, Brundin L. (2006) Neurogenesis in the adult spinal cord in an experimental model of multiple sclerosis. *Eur. J. Neurosci.* 23, 394-400.

Dash PK, Mach SA, Moore AN. (2001) Enhanced neurogenesis in the rodent hippocampus following traumatic brain injury. *J. Neurosci. Res.* 63, 313-9.

del Rio JA, Soriano E. (1989) Immunocytochemical detection of 5'-bromodeoxyuridine incorporation in the central nervous system of the mouse. *Brain Res. Dev. Brain Res.* 49, 311-7.

Delp MD, Armstrong RB, Godfrey DA, Laughlin MH, Ross CD, Wilkerson MK. (2001) Exercise increases blood flow to locomotor, vestibular, cardiorespiratory and visual regions of the brain in miniature swine. *J. Physiol.* 533, 849-59.

Doetsch F, Alvarez-Buylla A. (1996) Network of tangential pathways for neuronal migration in adult mammalian brain. *Proc. Natl. Acad. Sci. USA* 93, 14895-900.

Doetsch F, Garcia-Verdugo JM, Alvarez-Buylla A. (1997) Cellular composition and three-dimensional organization of the subventricular germinal zone in the adult mammalian brain. *J. Neurosci.* 17, 5046-61.

Doetsch F, Caille I, Lim DA, Garcia-Verdugo JM, Alvarez-Buylla A. (1999) Subventricular zone astrocytes are neural stem cells in the adult mammalian brain. *Cell* 97, 703-16.

Eadie BD, Redila VA, Christie BR. (2005) Voluntary exercise alters the cytoarchitecture of the adult dentate gyrus by increasing cellular proliferation, dendritic complexity, and spine density. *J. Comp. Neurol.* 486, 39-47.

Ekdahl CT, Mohapel P, Elmer E, Lindvall O. (2001) Caspase inhibitors increase short-term survival of progenitor-cell progeny in the adult rat dentate gyrus following status epilepticus. *Eur. J. Neurosci.* 14, 937-45.

Endl E, Gerdes J. (2000) The Ki-67 protein: fascinating forms and an unknown function. *Exp. Cell Res.* 257, 231-7.

Endo Y, Nishimura JI, Kobayashi S, Kimura F. (1997) Long-term glucocorticoid treatments decrease local cerebral blood flow in the rat hippocampus, in association with histological damage. *Neurosci.* 79, 745-52.

Engstrom Y, Eriksson S, Jildevik I, Skog S, Thelander L, Tribukait B. (19850 Cell cycle-dependent expression of mammalian ribonucleotide reductase. Differential regulation of the two subunits. *J. Biol. Chem.* 260, 9114-6.

Eriksson PS, Perfilieva E, Bjork-Eriksson T, Alborn AM, Nordborg C, Peterson DA, Gage FH. (1998) Neurogenesis in the adult human hippocampus. *Nat. Med.* 4, 1313-7.

Fanarraga ML, Avila J, Zabala JC. (1999) Expression of unphosphorylated class III beta-tubulin isotype in neuroepithelial cells demonstrates neuroblast commitment and differentiation. *Eur. J. Neurosci.* 11, 517-27.

Filippov V, Kronenberg G, Pivneva T, Reuter K, Steiner B, Wang LP, Yamaguchi M, Kettenmann H, Kempermann G. (2003) Subpopulation of nestin-expressing progenitor cells in the adult murine hippocampus shows electrophysiological and morphological characteristics of astrocytes. *Mol. Cell Neurosci.* 23, 373-82.

Florenes VA, Holm R, Myklebost O, Lendahl U, Fodstad O. (1994) Expression of the neuroectodermal intermediate filament nestin in human melanomas. *Cancer Res.* 54, 4-356.

Francis F, Koulakoff A, Boucher D, Chafey P, Schaar B, Vinet MC, Friocourt G, McDonnell N, Reiner O, Kahn A, McConnell SK, Berwald-Netter Y, Denoulet P, Chelly, J. (1999) Doublecortin is a developmentally regulated, microtubule-associated protein expressed in migrating and differentiating neurons. *Neuron* 23, 247-56.

Freese A, O'Rourke D, Judy K, O'Connor MJ. (1994) The application of 5-bromodeoxyuridine in the management of CNS tumors. *J. Neurooncol.* 20, 81-95.

Frielingsdorf H, Schwarz K, Brundin P, Mohapel P. (2004) No evidence for new dopaminergic neurons in the adult mammalian substantia nigra. *Proc. Natl. Acad. Sci. USA* 101, 10177-82.

Frisen J, Johansson CB, Torok C, Risling M, Lendahl U. (1995) Rapid, widespread, and longlasting induction of nestin contributes to the generation of glial scar tissue after CNS injury. *J. Cell Biol.* 131, 453-64.

Fukushima M, Barka T. (1976) The effects of 5-bromodeoxyuridine and isoproterenol on the postnatal differentiation of rat submandibular gland. *Am. J. Anat.* 147, 159-82.

Gavrieli Y, Sherman Y, Ben-Sasson SA. (1992) Identification of programmed cell death in situ via specific labeling of nuclear DNA fragmentation. *J. Cell Biol.* 119, 493-501.

Gage FH, Coates PW, Palmer TD, Kuhn HG, Fisher LJ, Suhonen JO, Peterson DA, Suhr, ST, Ray J. (1995) Survival and differentiation of adult neuronal progenitor cells transplanted to the adult brain. *Proc. Natl. Acad. Sci. USA* 92, 11879-83.

Gage FH. (2000) Mammalian neural stem cells. *Science* 287, 1433-8.

Garcia-Verdugo JM, Doetsch F, Wichterle H, Lim DA, Alvarez-Buylla. (1998) Architecture and cell types of the adult subventricular zone: in search of the stem cells. *J. Neurobiol.* 36, 234-48.

Gleeson JG, Lin PT, Flanagan LA, Walsh CA. (1999) Doublecortin is a microtubule-associated protein and is expressed widely by migrating neurons. *Neuron* 23, 257-71.

Gobbel GT, Bellinzona M, Vogt AR, Gupta N, Fike JR, Chan PH. (1998) Response ofpostmitotic neurons to X-irradiation: implications for the role of DNA damage inneuronal apoptosis. *J. Neurosci.* 18, 147-55.

Gould E, Tanapat P, McEwen BS, Flugge G, Fuchs E. (1998) Proliferation of granule cell precursors in the dentate gyrus of adult monkeys is diminished by stress. *Proc. Natl. Acad. Sci. USA* 95, 3168-71.

Gould E, Reeves AJ, Graziano MS, Gross CG. (1999a) Neurogenesis in the neocortex of adult primates. *Science* 286, 548-52.

Gould E, Reeves AJ, Fallah M, Tanapat P, Gross CG, Fuchs E. (1999b) Hippocampal neurogenesis in adult Old World primates. *Proc. Natl. Acad. Sci. USA* 96, 5263-7.

Gould E, Vail N, Wagers M, Gross CG. (2001) Adult-generated hippocampal and neocortical neurons in macaques have a transient existence. *Proc. Natl. Acad. Sci. USA* 98, 10910-7.

Gould E, Gross CG. (2002) Neurogenesis in adult mammals: some progress and problems. *J. Neurosci.* 22, 619-23.

Graham V, Khudyakov J, Ellis P, Pevny L. (2003) SOX2 functions to maintain neural progenitor identity. *Neuron* 39, 749-65.

Gratzner HG. (1982) Monoclonal antibody to 5-bromo- and 5-iododeoxyuridine: A new reagent for detection of DNA replication. *Science* 218, 474-5.

Hall PA, McKee PH, Menage HD, Dover R, Lane DP. (1993) High levels of p53 protein in UV-irradiated normal human skin. *Oncogene* 8, 203-7.

Hastings NB, Gould E. (1999) Rapid extension of axons into the CA3 region by adult-generated granule cells. *J. Comp. Neurol.* 413, 146-54. Erratum in: (1999) *J. Comp. Neurol.* 415, 144.

Haughey NJ, Nath A, Chan SL, Borchard AC, Rao MS, Mattson MP. (2002) Disruption of neurogenesis by amyloid beta-peptide, and perturbed neural progenitor cell homeostasis, in models of Alzheimer's disease. *J. Neurochem.* 6, 1509-24.

Hayes NL, Nowakowski RS. (2000) Exploiting the dynamics of S-phase tracers in developing brain: interkinetic nuclear migration for cells entering versus leaving the S-phase. *Dev. Neurosci.* 22, 44-55.

Hayes NL, Nowakowski RS. (2002) Dynamics of cell proliferation in the adult dentate gyrus of two inbred strains of mice. *Brain Res. Dev. Brain Res.* 134, 77-85.

Hendzel MJ, Wei Y, Mancini MA, Van Hooser A, Ranalli T, Brinkley BR, Bazett-Jones DP, Allis CD. (1997) Mitosis-specific phosphorylation of histone H3 initiates primarily within pericentromeric heterochromatin during G2 and spreads in an ordered fashion coincident with mitotic chromosome condensation. *Chromosoma* 106, 348-60.

Herrup K, Arendt T. (2002) Re-expression of cell cycle proteins induces neuronal cell death during Alzheimer's disease. *J. Alzheimers Dis.* 4, 243-7.

Herrup K, Neve R, Ackerman SL, Copani A. (2004) Divide and die: cell cycle events as triggers of nerve cell death. *J. Neurosci.* 24, 9232-9.

Hoshino T, Nagashima T, Cho KG, Davis RL, Donegan J, Slusarz M, Wilson CB. (1989) Variability in the proliferative potential of human gliomas. *J. Neurooncol.* 7, 137-43.

Hume WJ, Saffhill R. (1986) Iodo- and bromodeoxyuridine are excised at different ratesfrom DNA of mouse tongue keratinocytes in vitro. *Chem. Biol. Interact.* 60, 227-32.

Ide K, Secher NH. (2000) Cerebral blood flow and metabolism during exercise. *ProgNeurobiol.* 61, 397-414.

Ignatova TN, Kukekov VG, Laywell ED, Suslov ON, Vrionis FD, Steindler DA. (2002) Human cortical glial tumors contain neural stem-like cells expressing astroglialand neuronal markers in vitro. *Glia.* 39, 193-206.

Ino H, Chiba T. (2000) Expression of proliferating cell nuclear antigen (PCNA) in theadult and developing mouse nervous system. *Brain Res. Mol. Brain Res.* 78, 163-74.

Jin K, Sun Y, Xie L, Peel A, Mao XO, Batteur S, Greenberg DA. (2003) Directed migration of neuronal precursors into the ischemic cerebral cortex and striatum. *Mol. Cell Neurosci.* 24, 171-89.

Jin K, Peel AL, Mao XO, Xie L, Cottrell BA, Henshall DC, Greenberg DA. (2004a) Increased hippocampal neurogenesis in Alzheimer's disease. *Proc. Natl. Acad. Sci. USA* 101, 343-7.

Jin K, Galvan V, Xie L, Mao XO, Gorostiza OF, Bredesen DE, Greenberg DA. (2004b) Enhanced neurogenesis in Alzheimer's disease transgenic (PDGF-APPSw,I d) mice. *Proc. Natl. Acad. Sci. USA* 101, 13363-7.

Jin K, Xie L, Mao XO, Greenberg DA. (2006) Alzheimer's disease drugs promote neurogenesis. *Brain Res.* 1085, 183-8.

Kaneko Y, Sakakibara S, Imai T, Suzuki A, Nakamura Y, Sawamoto K, Ogawa Y, Toyama Y, Miyata T, Okano H. (2000) Musashi1: an evolutionally conserved marker for CNS progenitor cells including neural stem cells. *Dev. Neurosci.* 22, 139-53.

Kanemura Y, Mori K, Sakakibara S, Fujikawa H, Hayashi H, Nakano A, Matsumoto T, Tamura K, Imai T, Ohnishi T, Fushiki S, Nakamura Y, Yamasaki M, Okano H, Arita N, (2001) Musashi1, an evolutionarily conserved neural RNA-binding protein, is a versatile marker of human glioma cells in determining their cellular origin, malignancy, and proliferative activity. *Differentiation* 68, 141-52.

Kaplan MS, Hinds JW. (1977) Neurogenesis in the adult rat: electron microscopic analysis of light radioautographs. *Science* 197, 1092-4.

Kaplan MS. (1983) Proliferation of subependymal cells in the adult primate CNS: differential uptake of DNA labelled precursors. *J. Hirnforsch.* 24, 23-33.

Katchanov J, Harms C, Gertz K, Hauck L, Waeber C, Hirt L, Priller J, von Harsdorf R, Bruck W, Hortnagl H, Dirnagl U, Bhide PG, Endres M. (2001) Mild cerebral ischemia induces loss of cyclin-dependent kinase inhibitors and activation of cell cycle machinery before delayed neuronal cell death. *J. Neurosci.* 21, 5045-53.

Kato T, Yokouchi K, Fukushima N, Kawagishi K, Li Z, Moriizumi T. (2001) Continual replacement of newly-generated olfactory neurons in adult rats. *Neurosci. Lett.* 307, 17-20.

Katsetos CD, Del Valle L, Geddes JF, Assimakopoulou M, Legido A, Boyd JC, Balin B, Parikh NA, Maraziotis T, de Chadarevian JP, Varakis JN, Matsas R, Spano A, Frankfurter A, Herman MM, Khalili K. (2001) Aberrant localization of the neuronal class III beta-tubulin in astrocytomas. *Arch Pathol. Lab. Med.* 125, 613-24.

Katsetos CD, Herman MM, Mork SJ. (2003) Class III beta-tubulin in human development and cancer. *Cell Motil. Cytoskeleton.* 55, 77-96.

Kee N, Sivalingam S, Boonstra R, Wojtowicz JM. (2002) The utility of Ki-67 and BrdU as proliferative markers of adult neurogenesis. *J. Neurosci. Methods* 115, 97-105.

Kempermann G, Kuhn HG, Gage FH. (1997) More hippocampal neurons in adult mice living in an enriched environment. *Nature* 386, 493-5.

Keoffler HP, Yen J, Carlson J. (1983) The study of human myeloid differentiation using bromodeoxyuridine (BrdU). *J. Cell Physiol.* 116, 111-7.

Klein JA, Longo-Guess CM, Rossmann MP, Seburn KL, Hurd RE, Frankel WN, Bronson RT, Ackerman SL. (2002) The harlequin mouse mutation downregulates apoptosis-inducing factor. *Nature* 419, 367-74.

Klein JA, Ackerman SL. (2003) Oxidative stress, cell cycle, and neurodegeneration. *J. Clin. Invest.* 111, 785-93.

Koketsu D, Mikami A, Miyamoto Y, Hisatsune T. (2003) Nonrenewal of neurons in the cerebral neocortex of adult macaque monkeys. *J. Neurosci.* 23, 937-42.

Kokoeva MV, Yin H, Flier JS. (2005) Neurogenesis in the hypothalamus of adult mice: potential role in energy balance. *Science* 310, 679-83.

Kolb B, Pedersen B, Ballermann M, Gibb R, Whishaw IQ. (1999) Embryonic and postnatal injections of bromodeoxyuridine produce age-dependent morphological and behavioral abnormalities. *J. Neurosci.* 19, 2337-46.

Komitova M, Eriksson PS. (2004) Sox-2 is expressed by neural progenitors and astroglia in the adult rat brain. *Neurosci. Lett.* 369, 24-7.

Kopen GC, Prockop DJ, Phinney DG. (1999) Marrow stromal cells migrate throughout forebrain and cerebellum, and they differentiate into astrocytes after injection into neonatal mouse brains. *Proc. Natl. Acad. Sci. USA* 96, 10711-6.

Kornack DR, Rakic P. (1999) Continuation of neurogenesis in the hippocampus of the adult macaque monkey. *Proc. Natl. Acad. Sci. USA* 96, 5768-73.

Kornack DR, Rakic P. (2001a) The generation, migration, and differentiation of olfactory neurons in the adult primate brain. *Proc. Natl. Acad. Sci. USA* 98, 4752-7.

Kornack DR, Rakic P. (2001b) Cell proliferation without neurogenesis in adult primate neocortex. *Science* 294, 2127-30.

Korr H, Koeser K, Oldenkott S, Schmidt H, Schultze B. (1989) X-ray dose-effect relationship on unscheduled DNA synthesis and spontaneous unscheduled DNA synthesis in mouse brain cells studied in vivo. *Radiat. Environ. Biophys.* 28, 13-26.

Kriss JP, Maruyama Y, Tung LA, Bond SB, Revesz L. (1963) The fate of 5-bromodeoxyuridine, 5-bromodeoxycytidine, and 5-iododeoxycytidine in man. *Cancer Res.* 23, 260-8.

Kuan CY, Schloemer AJ, Lu A, Burns KA, Weng WL, Williams MT, Strauss KI, Vorhees CV, Flavell RA, Davis RJ, Sharp FR, Rakic P. (2004) Hypoxia-ischemia induces DNA synthesis without cell proliferation in dying neurons in adult rodent brain. *J. Neurosci.* 24, 10763-72.

Kuhn HG, Dickinson-Anson H, Gage FH. (1996) Neurogenesis in the dentate gyrus of the adult rat: age-related decrease of neuronal progenitor proliferation. *J. Neurosci.* 16, 2027-33.

Kuhn HG, Winkler J, Kempermann G, Thal LJ. Gage FH. (1997) Epidermal growth factor and fibroblast growth factor-2 have different effects on neural progenitors in the adult rat brain. *J. Neurosci.* 17, 5820-9.

Kurki P, Vanderlaan M, Dolbeare F, Gray J, Tan, E.M. (1986) Expression of proliferating cell nuclear antigen (PCNA)/cyclin during the cell cycle. *Exp. Cell Res.* 166, 209- 19.

Kuwagata M, Nagao T. (1998) Behavior and reproductive function of rat male offspring treated prenatally with 5-bromo-2'-deoxyuridine. *Reprod. Toxicol.* 12, 541-9.

Langley B, Ratan RR. (2004) Oxidative stress-induced death in the nervous system: cell cycle dependent or independent? *J. Neurosci. Res.* 77, 621-9.

Lendahl U, Zimmerman LB, McKay RD. (1990) CNS stem cells express a new class of intermediate filament protein. *Cell* 60, 585-95.

Li L, Story M, Legerski RJ. (2001a) Cellular responses to ionizing radiation damage. *Int. J. Radiat. Oncol. Biol. Phys.* 49, 1157-62.

Li Y, Chen J, Chopp M. (2001b) Adult bone marrow transplantation after stroke in adult rats. *Cell Transplant.* 10, 31-40.

Lie DC, Dziewczapolski G, Willhoite AR, Kaspar BK, Shults CW, Gage FH. (2002) The adult substantia nigra contains progenitor cells with neurogenic potential. *J. Neurosci.* 22, 6639-49.

Liu J, Solway K, Messing RO, Sharp FR. (1998) Increased neurogenesis in the dentate gyrus after transient global ischemia in gerbils. *J. Neurosci.* 18, 7768-78.

Liu DX, Greene LA. (2001) Neuronal apoptosis at the G1/S cell cycle checkpoint. *Cell Tissue Res.* 305, 217-28.

Lois C, Alvarez-Buylla A. (1994) Long-distance neuronal migration in the adult mammalian brain. *Science* 264, 1145-8.

Lois C, Garcia-Verdugo JM, Alvarez-Buylla A. (1996) Chain migration of neuronal precursors. *Science* 271, 978-81.

Lopez F, Belloc F, Lacombe F, Dumain P, Reiffers J, Bernard P, Boisseau MR. (1991) Modalities of synthesis of Ki-67 antigen during the stimulation of lymphocytes. *Cytometry* 12, 42-9.

Lowndes NF, Murguia JR. (2000) Sensing and responding to DNA damage. *Curr. Opin. Genet Dev.* 10, 17-25.

Lynch TP, Cass CE, Paterson AR. (1977) Defective transport of thymidine by cultured cells resistant to 5-bromodeoxyuridine. *J. Supramol. Struct.* 6, 363-74.

Magavi SS, Leavitt BR, Macklis JD. (2000) Induction of neurogenesis in the neocortex of adult mice. *Nature* 405, 951-5.

Malberg JE, Eisch AJ, Nestler EJ, Duman, RS. (2000) Chronic antidepressant treatment increases neurogenesis in adult rat hippocampus. *J. Neurosci.* 20, 9104-10.

Mann GJ, Musgrove EA, Fox RM, Thelander L. (1988) Ribonucleotide reductase M1 subunit in cellular proliferation, quiescence, and differentiation. *Cancer Res.* 48, 5151-6.

Markakis EA, Gage FH. (1999) Adult-generated neurons in the dentate gyrus send axonal projections to field CA3 and are surrounded by synaptic vesicles. *J. Comp. Neurol.* 406, 449-60.

Marusich MF, Furneaux HM, Henion PD, Weston JA. (1994) Hu neuronal proteins are expressed in proliferating neurogenic cells. *J. Neurobiol.* 25, 143-55.

Memo M. (1999) DNA damage and DNA repair in the nervous system. *Neurobiol. Aging* 20, 89-90.

Mezey E. (2004) Commentary: on bone marrow stem cells and openmindedness. *Stem Cells Dev.* 13, 147-52.

Miller MW, Nowakowski RS. (1988) Use of bromodeoxyuridine-immunohistochemistry to examine the proliferation, migration and time of origin of cells in the central nervous system. *Brain Res.* 457, 44-52.

Mizumatsu S, Monje ML, Morhardt DR, Rola R, Palmer TD, Fike JR. (2003) Extreme sensitivity of adult neurogenesis to low doses of X-irradiation. *Cancer Res.* 63, 4021-7.

Moran R, Darzynkiewicz Z, Staiano-Coico L, Melamed MR. (1985) Detection of 5-bromodeoxyuridine (BrdUrd) incorporation by monoclonal antibodies: role of the DNA denaturation step. *J. Histochem. Cytochem.* 33, 821-7.

Morris SM. (1991) The genetic toxicology of 5-bromodeoxyuridine in mammalian cells. *Mutat. Res.* 258, 161-88.

Morris SM, Domon OE, McGarrity LJ, Kodell RL, Casciano DA. (1992) Effect of bromodeoxyuridine on the proliferation and growth of ethyl methanesulfonate-exposed P3 cells: relationship to the induction of sister-chromatid exchanges. *Cell Biol. Toxicol.* 8, 75-87.

Morshead CM, van der Kooy D. (1992) Postmitotic death is the fate of constitutively proliferating cells in the subependymal layer of the adult mouse brain. *J. Neurosci.* 12, 249-56.

Mullen RJ, Buck CR, Smith AM. (1992) NeuN, a neuronal specific nuclear protein in vertebrates. *Development* 116, 201-11.

Nagai A, Matsumiya H, Yasui S, Aoshiba K, Ishihara Y, Konno K. (1993) Administration of bromodeoxyuridine in early postnatal rats results in lung changes at maturity. *Exp. Lung Res.* 19, 203-19.

Nagao T, Kuwagata M, Saito Y. (1998) Effects of prenatal exposure to 5-fluoro-2'-deoxyuridine on developing central nervous system and reproductive function in male offspring of mice. *Teratog. Carcinog. Mutagen* 18, 73-92.

Nakashima K, Ninomiya H, Fujiki Y. (1984) The effect of 5-bromodeoxyuridine on mouse embryos during neurulation in vitro. *Experientia* 40, 924-9.

Namura S, Zhu J, Fink K, Endres M, Srinivasan A, Tomaselli KJ, Yuan J, Moskowitz MA. (1998) Activation and cleavage of caspase-3 in apoptosis induced by experimental cerebral ischemia. *J. Neurosci.* 18, 3659-68.

Nowakowski RS, Lewin SB, Miller MW. (1989) Bromodeoxyuridine immunohistochemical determination of the lengths of the cell cycle and the DNA-synthetic phase for an anatomically defined population. *J. Neurocytol.* 18, 311-8.

Nowakowski RS, Hayes NL. (2000) New neurons: extraordinary evidence or extraordinary conclusion? *Science* 288, 771.

Nowakowski RS, Hayes N.L. (2001) Stem cells: the promises and pitfalls Neuropsychopharmacology. 25, 799-804.

Okuda T, Tagawa K, Qi ML, Hoshio M, Ueda H, Kawano H, Kanazawa I, Muramatsu M, Okazawa H. (2004) Oct-3/4 repression accelerates differentiation of neural progenitor cells in vitro and in vivo. *Brain Res. Mol. Brain Res.* 132, 18-30.

Oltra S, Martinez F, Orellana C, Grau E, Fernandez JM, Canete A, Castel V. (2005) The doublecortin gene, a new molecular marker to detect minimal residual disease in neuroblastoma. *Diagn. Mol. Pathol.* 14, 53-7.

Packard DS Jr, Menzies RA, Skalko RG. (1973) Incorportaiton of thymidine and its analogue, bromodeoxyuridine, into embryos and maternal tissues of the mouse. *Differentiation* 1, 397-404.

Palmer TD, Willhoite AR, Gage FH. (2000) Vascular niche for adult hippocampal neurogenesis. *J. Comp. Neurol.* 425, 479-94.

Parent JM, Yu TW, Leibowitz RT, Geschwind DH, Sloviter RS, Lowenstein D.H. (1997) Dentate granule cell neurogenesis is increased by seizures and contributes to aberrant network reorganization in the adult rat hippocampus. *J. Neurosci.* 17, 3727-38.

Parent JM, Tada E, Fike JR, Lowenstein DH. (1999) Inhibition of dentate granule cell neurogenesis with brain irradiation does not prevent seizure-induced mossy fiber synaptic reorganization in the rat. *J. Neurosci.* 19, 4508-19.

Pencea V, Bingaman KD, Freedman LJ, Luskin MB. (2001) Neurogenesis in the subventricular zone and rostral migratory stream of the neonatal and adult primate forebrain. *Exp. Neurol.* 172, 1-16.

Pompeiano M, Blaschke AJ, Flavell RA, Srinivasan A, Chun J. (2000) Decreased apoptosis in proliferative and postmitotic regions of the Caspase 3-deficient embryonic central nervous system. *J. Comp. Neurol.* 423, 1-12.

Pont F, Collet A, Lallement G. (1995) Early and transient increase of rat hippocampal blood-brain barrier permeability to amino acids during kainic acid-induced seizures. *Neurosci Lett.* 184, 52-4.

Qu TY, Dong XJ, Sugaya I, Vaghani A, Pulido J, Sugaya K. (2004) Bromodeoxyuridine increases multipotency of human bone marrow-derived stem cells. *Restor Neurol Neurosci.* 22, 459-68.

Quinn CC, Gray GE, Hockfield S. (1999) A family of proteins implicated in axon guidance and outgrowth. *J. Neurobiol.* 41, 158-64.

Rakic P. (1974) Neurons in rhesus monkey visual cortex: systematic relation between time of origin and eventual disposition. *Science* 183, 425-7.

Rakic P. (1985) Limits of neurogenesis in primates. *Science* 227, 1054-6.

Rakic P. (2002) Adult neurogenesis in mammals: an identity crisis. *J. Neurosci.* 22, 614-8.

Rehen SK, McConnell MJ, Kaushal D, Kingsbury MA, Yang AH, Chun J. (2001) Chromosomal variation in neurons of the developing and adult mammalian nervous system. *Proc. Natl. Acad. Sci. USA* 98, 13361-6.

Rehen SK, Yung YC, McCreight MP, Kaushal D, Yang AH, Almeida BS, Kingsbury MA, Cabral KM, McConnell MJ, Anliker B, Fontanoz M, Chun J. (2005) Constitutional aneuploidy in the normal human brain. *J. Neurosci.* 25, 2176-80.

Rietze R, Poulin P, Weiss S. (2000) Mitotically active cells that generate neurons and astrocytes are present in multiple regions of the adult mouse hippocampus. *J. Comp. Neurol.* 424, 397-408.

Ribatti D, Nico B, Crivellato E, Artico M. (2006) Development of the blood-brain barrier: a historical point of view. Anat. Rec. B *New Anat.* 289, 3-8.

Rousselot P, Lois C, Alvarez-Buylla A. (1995) Embryonic (PSA) N-CAM reveals chains of migrating neuroblasts between the lateral ventricle and the olfactory bulb of adult mice. *J. Comp. Neurol.* 351, 51-61.

Saffhill R, Ockey CH (1985) Strand breaks arising from the repair of the 5-bromodeoxyuridine-substituted template and methyl methanesulphonate-induced lesions can explain the formation of sister chromatid exchanges. *Chromosoma* 92, 218-24.

Sanai N, Tramontin AD, Quinones-Hinojosa A, Barbaro NM, Gupta N, Kunwar S, Lawton MT, McDermott MW, Parsa AT, Garcia-Verdugo JM, Berger MS, Alvarez-Buylla A. (2004) Unique astrocyte ribbon in adult human brain contains neural stem cells but lacks chain migration. *Nature* 427, 740-4.

Santarelli L, Saxe M, Gross C, Surget A, Battaglia F, Dulawa S, Weisstaub N, Lee J, Duman R, Arancio O, Belzung C, Hen R. (2003) Requirement of hippocampal neurogenesis for the behavioral effects of antidepressants. *Science* 301, 805-9.

Schlessinger AR, Cowan WM, Gottlieb DI. (1975) An autoradiographic study of the time of origin and the pattern of granule cell migration in the dentate gyrus of the rat. *J. Comp. Neurol.* 159, 149-75.

Schmitz C, Axmacher B, Zunker U, Korr H. (1999) Age-related changes of DNA repair and mitochondrial DNA synthesis in the mouse brain. *Acta. Neuropathol.* (Berl). 97, 71-81.

Scholzen T, Gerdes J. (2000) The Ki-67 protein: from the known and the unknown. *J. Cell Physiol.* 182, 311-22.

Sekerkova G, Ilijic E, Mugnaini E. (2004) Bromodeoxyuridine administered during neurogenesis of the projection neurons causes cerebellar defects in rat. *J. Comp. Neurol.* 470, 221-39.

Seki T, Arai Y. (1993) Highly polysialylated neural cell adhesion molecule (NCAM-H) is expressed by newly generated granule cells in the dentate gyrus of the adult rat. *J. Neurosci.* 13, 2351-8.

Seki T, Arai Y. (1999) Temporal and spacial relationships between PSA-NCAM- expressing, newly generated granule cells, and radial glia-like cells in the adult dentate gyrus. *J. Comp. Neurol.* 410, 503-13.

Selden JR, Dolbeare F, Clair JH, Nichols WW, Miller JE, Kleemeyer KM, Hyland RJ, DeLuca JG. (1993) Statistical confirmation that immunofluorescent detection of DNA repair in human fibroblasts by measurement of bromodeoxyuridine incorporation is stoichiometric and sensitive. *Cytometry* 14, 154-67.

Selden JR, Dolbeare F, Clair JH, Miller JE, McGettigan K, DiJohn JA, Dysart GR, DeLuca JG. (1994) Validation of a flow cytometric in vitro DNA repair (UDS) assay in rat hepatocytes. *Mutat. Res.* 315, 147-67.

Seri B, Garcia-Verdugo JM, McEwen BS, Alvarez-Buylla A. (2001) Astrocytes give rise to new neurons in the adult mammalian hippocampus. *J. Neurosci.* 21, 7153-60.

Sharma HS, Cervos-Navarro J, Dey PK. (1991) Increased blood-brain barrier permeability following acute short-term swimming exercise in conscious normotensive young rats. *Neurosci. Res.* 10, 211-21.

Shihabuddin LS, Horner PJ, Ray J, Gage FH. (2000) Adult spinal cord stem cells generate neurons after transplantation in the adult dentate gyrus. *J. Neurosci.* 20, 8727-35.

Shinohara C, Gobbel GT, Lamborn KR, Tada E, Fike JR. (1997) Apoptosis in the subependyma of young adult rats after single and fractionated doses of X-rays. *Cancer Res.* 57, 2694-702.

Sidman RL, Miale IL, Feder N. (19590 Cell proliferation and migration in the primitive ependymal zone: an autoradiographic study of histogenesis in the nervous system. *Exp. Neurol.* 1, 322-33.

Sidman RL. (1970) Autoradiographic methods and principles for study of the nervous system with thymidine-H3. Contemporary Research Methods in Neuroanatomy. W.J.H. Nauta and S.O.E. Ebbesson. New York, Springer: 252-274.

Sloviter RS. (1989) Calcium-binding protein (calbindin-D28k) and parvalbumin immunocytochemistry: localization in the rat hippocampus with specific reference to the selective vulnerability of hippocampal neurons to seizure activity. *J. Comp. Neurol.* 280, 183-96.

Snider BJ, Gottron FJ, Choi DW. (1999) Apoptosis and necrosis in cerebrovascular disease. *Ann. N. Y. Acad. Sci.* 893, 243-53.

Spalding KL, Bhardwaj RD, Buchholz BA, Druid H, Frisen J. (2005) Retrospective birth dating of cells in humans. *Cell* 122, 133-43.

Spector R, Berlinger WG. (1982) Localization and mechanism of thymidine transport in the central nervous system. *J. Neurochem*. 39, 837-41.

Stanfield BB, Trice JE. (1988) Evidence that granule cells generated in the dentate gyrus of adult rats extend axonal projections. *Exp. Brain Res*. 72, 399-406.

Struikmans H, Rutgers DH, Jansen GH, Tulleken CA, van der Tweel I, Batterman JJ, (1997) S-phase fraction, 5-bromo-2'-deoxy-uridine labelling index, duration of S-phase, potential doubling time, and DNA index in benign and malignant brain tumors. *Radiat. Oncol. Investig.* 5, 170-9.

Suhonen JO, Peterson DA, Ray J, Gage FH. (1996) Differentiation of adult hippocampus-derived progenitors into olfactory neurons in vivo. *Nature* 383, 624-7.

Sun W, Winseck A, Vinsant S, Park OH, Kim H, Oppenheim RW. (2004) Programmed cell death of adult-generated hippocampal neurons is mediated by the proapoptotic gene Bax. *J. Neurosci.* 24, 11205-13.

Swain RA, Harris AB, Wiener EC, Dutka MV, Morris HD, Theien BE, Konda S, Engberg K, Lauterbur PC, Greenough WT. (2003) Prolonged exercise induces angiogenesis and increases cerebral blood volume in primary motor cortex of the rat. *Neurosci*. 117, 1037-46.

Tada E, Yang C, Gobbel GT, Lamborn KR, Fike JR. (1999) Long-term impairment of subependymal repopulation following damage by ionizing irradiation. *Exp. Neurol.* 160, 66-77.

Tada E, Parent JM, Lowenstein DH, Fike JR. (2000) X-irradiation causes a prolonged reduction in cell proliferation in the dentate gyrus of adult rats. *Neurosci.* 99, 33-41.

Tai MH, Chang CC, Kiupel M, Webster JD, Olson LK, Trosko JE. (2005) Oct4 expression in adult human stem cells: evidence in support of the stem cell theory of carcinogenesis. Carcinogenesis. 26, 495-502. Erratum in: (2005) *Carcinogenesis* 26, 1316.

Takahashi T, Caviness VS Jr. (1993) PCNA-binding to DNA at the G1/S transition in proliferating cells of the developing cerebral wall. *J. Neurocytol.* 22, 1096-102.

Takahashi T, Nowakowski RS, Caviness VS Jr. (1995) The cell cycle of the pseudostratified ventricular epithelium of the embryonic murine cerebral wall. *J. Neurosci*. 15, 6046-57.

Tanapat P, Hastings NB, Reeves AJ, Gould E. (1999) Estrogen stimulates a transient increase in the number of new neurons in the dentate gyrus of the adult female rat. *J. Neurosci.* 19, 5792-801.

Tattersfield AS, Croon RJ, Liu YW, Kells AP, Faull RL, Connor B. (2004) Neurogenesis in the striatum of the quinolinic acid lesion model of Huntington's disease. *Neurosci.* 127, 319-32.

Taupin P, Ray J, Fischer WH, Suhr ST, Hakansson K, Grubb A, Gage FH. (2000) FGF-2-responsive neural stem cell proliferation requires CCg, a novel autocrine/paracrine cofactor. *Neuron* 28, 385-97.

Taupin P, Gage FH. (2002) Adult neurogenesis and neural stem cells of the central nervous system in mammals. *J. Neurosci. Res.* 69, 745-9.

Taupin P. (2005) Adult neurogenesis in the mammalian central nervous system: functionality and potential clinical interest. *Med. Sci. Monit.* 11, RA247-52.

Taupin P. (2006) Neurogenesis in the adult central nervous system. *C. R. Biol.* 329, 65-475.

Thomas nee Williams SA, Segal MB. (1996) Identification of a saturable uptake system for deoxyribonucleosides at the blood-brain and blood-cerebrospinal fluid barriers. *Brain Res.* 741, 230-9.

Thomas SA, Davson H, Segal MB. (1997) Quantification of efflux into the blood and brain of intraventricularly perfused [3H]-thymidine in the anaesthetized rabbit. *Exp. Physiol.* 82, 139-48.

Thomas SA, Segal MB. (1997) Saturation kinetics, specificity and NBMPR sensitivity of thymidine entry into the central nervous system. *Brain Res.* 760, 59-67.

Tillfors M, Furmark T, Marteinsdottir I, Fischer H, Pissiota A, Langstrom B, Fredrikson M. (2001) Cerebral blood flow in subjects with social phobia during stressful speaking tasks: a PET study. *Am. J. Psychiatry* 158, 1220-6.

van Praag H, Kempermann G, Gage .H. (1999) Running increases cell proliferation and neurogenesis in the adult mouse dentate gyrus. *Nat. Neurosci.* 2, 266-70.

van Praag H, Kempermann G, Gage FH. (2000) Neural consequences of environmental enrichment. *Nat. Rev. Neurosci.* 1, 191-8.

van Praag H, Schinder AF, Christie BR, Toni N, Palmer TD, Gage FH. (2002) Functional neurogenesis in the adult hippocampus. *Nature* 415, 1030-4.

Vincent I, Rosado M, Davies P. (1996) Mitotic mechanisms in Alzheimer's disease? *J. Cell Biol.* 132, 413-25.

West MJ, Slomianka L, Gundersen HJ. (1991) Unbiased stereological estimation of the total number of neurons in thesubdivisions of the rat hippocampus using the optical fractionator. *Anat. Rec.* 231, 482-97.

Wichterle H, Garcia-Verdugo JM, Alvarez-Buylla A. (1997) Direct evidence for homotypic, glia-independent neuronal migration. *Neuron* 18, 779-91.

Wilt FH, Anderson M. (1972) The action of 5-bromodeoxyuridine on differentiation *Dev. Biol.* 28, 443-7.

Yamada M, Onodera M, Mizuno Y, Mochizuki H. (2004) Neurogenesis in olfactory bulb identified by retroviral labeling in normal and 1-methyl-4-phenyl-1,2,3,6-tetrahydropyridine-treated adult mice. *Neurosci.* 124, 173-81.

Yamaguchi M, Saito H, Suzuki M, Mori K. (2000) Visualization of neurogenesis in the central nervous system using nestin promoter-GFP transgenic mice. *Neuroreport* 11, 1991-6.

Yang Y, Geldmacher DS, Herrup K. (2001) DNA replication precedes neuronal cell death in Alzheimer's disease. *J. Neurosci.* 21, 2661-8.

Yang Y, Mufson EJ, Herrup K. (2003) Neuronal cell death is preceded by cell cycle events at all stages of Alzheimer's disease. *J. Neurosci.* 23, 2557-63.

Yu, W.H. (1976) The effect of 5-bromodeoxyuridine on the postnatal development of the rat cerebellum: a biochemical study. *Brain Res.* 118, 281-91.

Younkin L, Silberberg D. (1973) Myelination in developing cultured newborn rat cerebellum inhibited by 5-bromodeoxyuridine. *Exp. Cell Res.* 76, 455-8.

Younkin LH, Silberberg DH. (1976) Delay of oligodendrocyte differentiation by 5-bromodeoxyuridine (BUdR). *Brain Res.* 101, 600-5.

Zacchetti A, van Garderen E, Teske E, Nederbragt H, Dierendonck JH, Rutteman GR, (2003) Validation of the use of proliferation markers in canine neoplastic and non-neoplastic tissues: comparison of KI-67 and proliferating cell nuclear antigen (PCNA) expression versus in vivo bromodeoxyuridine labelling by immunohistochemistry. *APMIS* 111, 430-8.

Zhao M, Momma S, Delfani K, Carlen M, Cassidy RM, Johansson CB, Brismar H, Shupliakov O, Frisen J, Janson AM. (2003) Evidence for neurogenesis in the adult mammalian substantia nigra. *Proc. Natl. Acad. Sci. USA* 100, 7925-30.

Zhu H, Wang ZY, Hansson HA. (2003) Visualization of proliferating cells in the adult mammalian brain with the aid of ribonucleotide reductase. *Brain Res*. 977, 180-9.

Zhu H, Dahlstrom A, Hansson HA. (2005) Characterization of cell proliferation in the adult dentate under normal conditions and after kainate induced seizures using ribonucleotide reductase and BrdU. *Brain Res.* 1036, 7-17.

Chapter III

Adult Neural Stem Cells, Neurogenic Niches and Cellular Therapy

Abstract

Niches are specialized microenvironments that regulate stem cells activity. In the nervous system, during development, niches control neural stem cells (NSCs) maturation and the formation of the neuronal network. In the adult, neurogenesis occurs in discrete areas of the brain, the subventricular zone (SVZ) and the hippocampus, where neurogenic niches have been identified and characterized. These niches, an angiogenic and an astroglial niches, control NSCs self-renewal and differentiation. Though the molecular and cellular mechanisms underlying the interactions between NSCs and their environment remain to be elucidated, neurogenic niches share similar developmentally conserved pathways with other niches. It is hypothesized that neurogenic niches underlie the properties and functions of NSCs in the adult central nervous system (CNS). Hence, neurogenic niches may not only hold the key to our understanding of neurogenesis in the adult brain, but also of the developmental potential of adult NSCs, and their potential for cellular therapy.

Introduction

Stem cells reside in specialized microenvironments or "niches" that regulate their self-renewal and differentiation activities. Hence, the niches in which they reside regulate the behavior of stem cells. Niches have been identified and characterized during development, and in various adult tissues [1]. In the nervous system, it is hypothesized that niches control the developmental potential of NSCs. During development, the niches control the maturation of the CNS; the neurogenic niches are spatially and temporally regulated, as neurogenesis is unfolding. In contrast, in the adult CNS, as neurogenesis occurs in discrete areas of the brain, the neurogenic niches are maintained in restricted regions throughout adulthood [2].

Adult Neurogenesis and NSCs

Contrary to the long held dogma, neurogenesis occurs throughout adulthood in the mammalian brain [3]. Neurogenesis occurs primarily in two areas of the adult brain, the SVZ and dentate gyrus (DG) of the hippocampus, in various species including human [4]. In the DG, newly generated neuronal cells in the subgranular zone (SGZ) migrate to the granular layer, where they differentiate into mature neuronal cells, and extend axonal projections to the CA3 area [5, 6]. In the SVZ, cells are generated in the anterior part of the SVZ, and migrate as chains, tangentially along the SVZ and trough the rostro-migratory stream (RMS), to the olfactory bulb (OB) where they differentiate into interneurons [7-11]. Neurogenesis has been reported to occur in other areas of the adult brain in certain species, like the CA1 area, the neocortex, the striatum, the amygdala, the substantia nigra, the 3rd ventricle, the subcortical white matter, the caudate nucleus. However some of these data have been the source of debates and controversies, and remain to be further confirmed [4]. In the human brain, neurogenesis has been characterized in the hippocampus [12], but not in the SVZ [13]. Sanai et al. found no evidence of migrating chains of neuroblasts in the SVZ or in the pathway to the OB, suggesting that no new neuronal cells are generated from the SVZ in the adult human brain, contrary to other species including non-human primates [13]. Sanai et al. further reported that the human SVZ has a distinct organization not observed in other species, including non-human primates; it contains a ribbon of astrocytes. The authors hypothesized that such particular organization may underlie the lack of neurogenesis in the human SVZ [13]. Alternatively, precursor cells in the human SVZ may migrate as individual cells, or the pathway may be different and yet to be identified.

It is hypothesized that newly generated neuronal cells in the adult brain originate from NSCs [14]. Neural stem cells (NSCs) are the self-renewing, multipotent cells that generate neurons, astrocytes, and oligodendrocytes in the nervous system. Self-renewing, multipotent NSCs have been isolated and characterized *in vitro*, from the SVZ [15, 16] and hippocampus [17, 18], supporting the existence of NSCs in the adult brain. Recent reports have challenged the *in vitro* isolation and characterization of self-renewing, multipotent NSCs from the adult hippocampus, claiming that the adult hippocampus contains neural progenitor cells, but not self-renewing, multipotent NSCs [19, 20]. Neural progenitor cells are multipotent cells with limited self-renewing capacity, as opposed to NSCs that have unlimited self-renewing capacity. However, differences in protocols, species, handling technique may explain discrepancies between the studies [21]. Neural progenitor and stem cells have also been isolated and characterized *in vitro* form other areas of the CNS, including the spinal cord [4]. Altogether, these data suggest that neural progenitor and stem cells reside in the adult CNS.

There are two theories with regard to the origin of newly generated neuronal cells in the adult brain. On the one hand, NSCs of the adult SVZ are differentiated ependymal cells that express the intermediate filament protein nestin [22]. Nestin is an intermediate filament that has been characterized as a marker for neuroepithelial and CNS stem cells [23]. On the other hand, NSCs are astrocyte-like cells expressing glial fibrillary acidic protein and nestin in the SVZ and SGZ [24-27], a hypothesis that has gained further support more recently [28-30]. In the adult spinal cord, gliogenesis, but not neurogenesis, occurs throughout the cord [31]. It has been hypothesized that the central canal in the adult spinal cord is the presumed location of CNS progenitor cells, since cells in the corresponding region of the brain, the SVZ, can proliferate and differentiate into neurons and glial cells [24-30].

Horner et al. reported that cell division occurs throughout the adult spinal cord, and is not restricted to the lining of the central canal, with the majority of dividing cells residing in the outer circumference of the spinal cord [31]. Hence, glial progenitor cells exist also in the outer circumference of the spinal cord, and proliferate and differentiate throughout the adult spinal cord. Horner et al. proposed two models with regard to the origin of glial progenitor cells in the adult spinal cord [31]. One model contends that a stem cell exists at the ependymal layer area, and divides asymmetrically. A daughter cell then migrates to the outer circumference of the spinal cord where it exists as a bipotent or glial progenitor cell, and begins to divide more rapidly. The other model predicts that a glial progenitor and stem cell population may exist in the outer circumference of the spinal cord, where cell division is more common. This model functionally separates ependymal cell division from the proliferative zone of the outer annuli. However, in contrast to the adult brain, newly generated cells in the adult spinal cord give rise to new cells restricted to the glial phenotype [31].

Two hypotheses can be formulated to explain such discrepancy. First, the adult spinal cord, as opposed to the adult brain, does not contain NSCs, but restricted glial progenitor cells. Alternatively, the adult spinal cord would contain NSCs, but the environment would prevent these cells to differentiate into neuronal lineage. The later hypothesis is further supported by *ex vivo* studies performed in the adult rat CNS, where neural progenitor and stem cells isolated from the adult spinal cord -a non-neurogenic area- respond to different cues whether they are grafted in the spinal cord or in the DG. When grafted in the spinal cord, the neural progenitor and stem cells derived from the spinal cord differentiated into astrocytes, whereas when grafted in the DG, the neural progenitor and stem cells differentiated into neuronal cells, as well as astrocytes [32].

Hence, grafted neural progenitor and stem cells behave like endogenous proliferating spinal cord cells, by differentiating into glia only when grafted in the spinal cord [31]. The ability of the cells to differentiate into neuronal phenotype in heterotypic transplantation studies, suggests that adult spinal cord derived-neural progenitor and stem cells are induced to express mature neuronal phenotypes by environmental signals.

Neurogenic Niches

Stem cell niches have been identified and characterized in various adult tissues, such as in the skin, bone marrow, placenta, liver and gut where they regulate the fate of stem cells [33-40]. In the adult brain, several niches have been identified, and associated with NSCs. Neurogenesis has been associated with angiogenesis in the adult hippocampus, and endothelial cells release soluble factors that stimulate the self-renewal of NSCs, defining the angiogenic or vascular niche for neurogenesis [41, 42]. The association of angiogenesis and neurogenesis suggests a mechanism by which neurogenesis can respond and adapt to the general state of the body. Astrocytes derived from neonatal SVZ [43] and from adult hippocampus [44] stimulate the differentiation of adult-derived neural progenitor and stem cells toward the neuronal lineage in co-culture, defining the astrocytic niche for neurogenesis. In the adult brain, it has been reported that populations of astrocytes act as primary precursors for the new neurons [24-30]. Hence, astrocytes not only may represent NSCs in the adult brain, but they also participate to the microenvironment that promotes neurogenesis in the germinal layers. The involvement of endothelial cells and astrocytes in the neurogenic niches suggests that factors controlling angiogenesis and gliogenesis may also control neurogenesis. In support of this contention, vascular endothelial growth factor (VEGF) receptors, whose ligands contribute to vascular development [45] and stimulate neurogenesis [46], are observed in the vascular niche for neurogenesis [41]. Other candidates that contribute to the angiogenic niche for neurogenesis include the proteins of the Inhibitor of Differentiation/DNA binding (Id) family, which are members of the basic helix-loop-helix family of transcription factors, but are lacking DNA binding domain [47]. Among the factors involved in the astrocytic niche are the carbohydrate moiety Lewis X (LeX) and the Eph/ephrin family of signaling molecules. LeX is expressed in the SVZ and granular layer where it appears to be associated with a subpopulation of astrocytes *in vivo*, and *in vitro*, SVZ cells that express LeX elicit self-renewal, multipotent properties [48]. SVZ astrocytes have been reported to express ephrin b2/3 ligands [49]. These ligands may also be involved in the angiogenic niche for neurogenesis, as they have been reported to be critical for vasculogenesis [50]. Interestingly, cystatin C, previously reported as a NSC factor for neural progenitor and stem cells *in vitro* and *in vivo* [51] is expressed by glial cells [51, 52], and in blood vessels [53], further supporting its role in neurogenesis and in the neurogenic niches. Microglial cells have also been identified and characterized as part of the neurogenic niches, and transforming growth factor beta synthesized by activated microglia promotes neurogenesis [54]. It is further proposed that the balance between pro- and anti-inflammatory secreted molecules influences the activity of the microglial niche for neurogenesis [54].

Niches are ancient evolutionary structures with conserved features across diverse tissues and organisms [55]. In support of this contention, the involvement of classical developmental signals and morphogens, like Notch, bone morphogenetic proteins (BMPs), Eph/ephrins, Noggin, Wint (Wnt) signaling molecules and Sonic hedgehog (Shh) have been reported in various niches across various species [33-40]. Recently, BMPs, Noggin, Shh and Wnt have been involved in adult neurogenesis [56-59], supporting their role in the neurogenic niches. The BMP family instructs adult NSCs to adopt a glial fate [56]. Antagonizing BMP signaling causes neurogenesin1, a secreted factor from astrocytes, to promote neurogenesis [60]. Noggin, also a BMP antagonist, and Wnt signaling play key roles in promoting neurogenesis in the adult hippocampus [57, 59]. Niches also display important differences in their organization that may underlie their specificities and functionalities [61].

During development, gliogenesis follows neurogenesis, whereas in the adult, the newly generated neuronal cells integrate a pre-existing network, where the glial cells are already in place [44, 62]. Adult spinal cord-derived neural progenitor and stem cells are induced to express mature neuronal phenotypes by environmental signals [32]. Contrary to astrocytes from adult hippocampus that are capable of regulating neurogenesis by instructing the stem cells to adopt a neuronal fate, astrocytes from adult cord are ineffective in promoting neurogenesis from adult stem cells [44]. These data further emphasize the role of the microenvironment in regulating the fate of neural progenitor and stem cells, and point to specific properties of the neurogenic niches during development and in the adult. In the postnatal SVZ, gamma-aminobutyric acid (GABA) signaling between neuroblasts and astrocytes limits stem cell proliferation and therefore, may contribute to maintaining a balance between the amplification and mobilization of progenitors; the neuroblasts release the neurotransmitter GABA that regulate the division of GFAP-expressing progenitor/stem cells, providing a feedback mechanism to control the proliferation of stem cells [63]. Altogether these data show that adult NSC fate specification is under a complex, yet stringent, control of a multitude of molecular signals, both instructive and permissive, that specifically instruct NSCs in their respective niches [64, 65]. It has also been reported that NSCs, under certain conditions, give rise to endothelial cells [66]. Thus, not only, stem cell niches control the developmental potential of NSCs, but NSCs may also contribute to create their own environments.

Broader Potential of Adult Stem Cells

Contrary to embryonic stem cells, the archetype of pluripotent stem cells, adult stem cells are multipotents; they generate lineage specific cell types restricted to the tissues from which they are derived. Several studies have reported data that adult-derived stem cells may have a broader potential; they may generate cell types of tissues other than the ones they are derived (figure 1). Genetically marked adult-derived clonal neurospheres were reported to give rise to blood cells upon transplantation into irradiated mice [67], skeletal myotubes on co-culture with a myogenic cell line or transplantation into regenerating muscle [68], and to contribute to tissues from all three germ layers upon transplantation into blastocysts [69]. Other studies reported that a homogenous population of adult-derived NSCs, purified by flow cytometry, differentiates into myocytes *in vitro* [70], that adult OB-derived neurospheres in co-culture with skeletal myoblasts generate skeletal muscle cells and give rise to myogenic progenitor cells that form myotubes *de novo* [71]. The ability to differentiate into lineages others than the tissues from where they originate has also been reported for stem cells derived from other tissues than the brain. Progenitor cells isolated from adult muscle tissue gave rise to blood cells upon transplantation into irradiated mice [72, 73], and neuronal cell lineages *in vitro* [74]. Adult bone marrow progenitor cells and purified hematopoietic stem cells, either genetically marked or identifiable by Y/X chromosome discrimination, were observed upon transplantation to give rise to skeletal myotubes and myocytes [75-77], hepatocytes [78-80], neurons and glial cells after transplantation [81-85]. Adult skin stem cells have been reported to differentiate into neurons, glial cells, smooth muscle cells and adipocytes *in vitro* [86, 87]. These data show that adult-derived stem cells might not be restricted to generate tissue specific cell types, but appear to have a wider differentiation potential than previously thought [88]. Though some of these observations have been attributed to phenomenon such as artifacts, contamination, transformation, transdifferentiation or cell fusion, certain reports presented suggest pluripotentiality as possible explanation for the phenotypes observed in these studies [89-93]. The evidences that adult-derived stem cells have a broader potential not only challenge the concept of multipotentiality, but also suggest that alternative sources of adult stem cells may be use for cellular therapy, particularly for the CNS.

It is hypothesized that stem cell niches may hold the clues of such stem cell plasticity; stem cells transplanted heterotopically would adopt the fate of the host-niches. The mechanisms of such plasticity remain to be elucidated. It is postulated that though niches are developmentally conserved, specific features of the niches may underlie such plasticity of adult stem cells. Removing a stem or progenitor cell from its original environment that regulates its fate, may lead to the regulation, reexpression, or *de novo* synthesis of molecules that would confer the stem cells a new fate, under a new environment [94]. An improved understanding of stem cell microenvironments will therefore provide clues on the potential of adult stem cells, and on how to send stem cells down to different developmental pathways. The broader potential of adult stem cells remains however the subject of debates and controversies [89-93]. Irrespectively of these controversies, the ability for adult stem and progenitor cells to adopt different phenotypes has tremendous therapeutic potential.

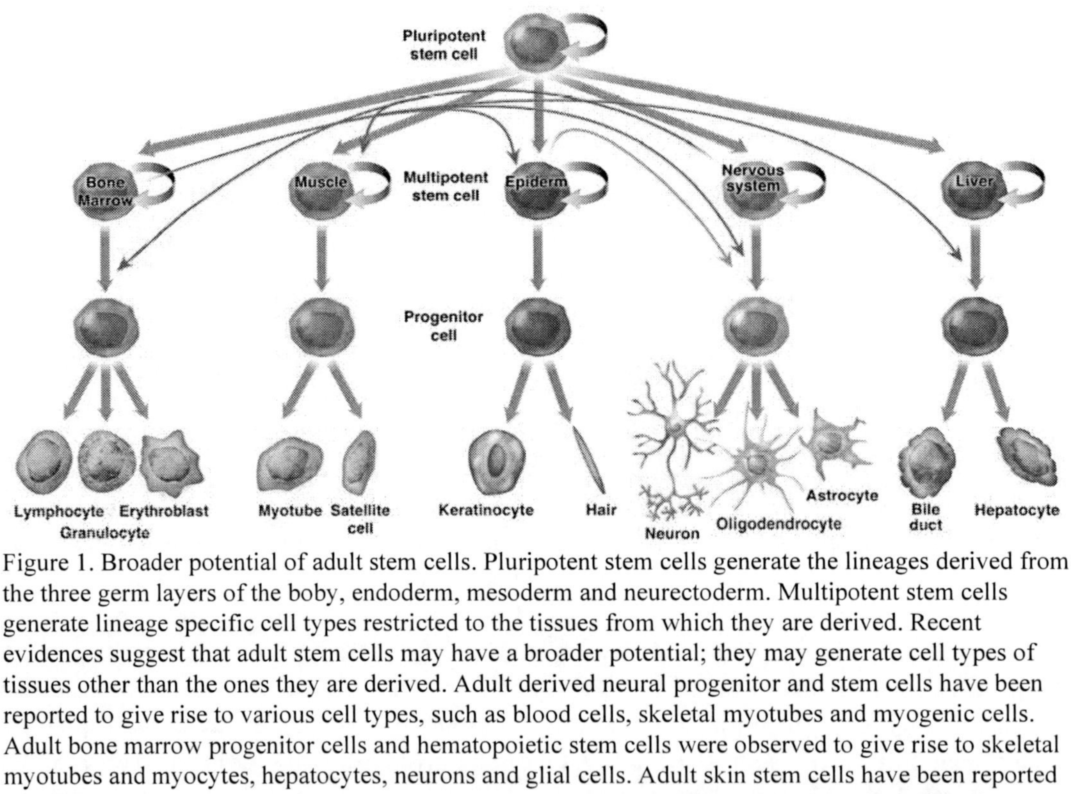

Figure 1. Broader potential of adult stem cells. Pluripotent stem cells generate the lineages derived from the three germ layers of the boby, endoderm, mesoderm and neurectoderm. Multipotent stem cells generate lineage specific cell types restricted to the tissues from which they are derived. Recent evidences suggest that adult stem cells may have a broader potential; they may generate cell types of tissues other than the ones they are derived. Adult derived neural progenitor and stem cells have been reported to give rise to various cell types, such as blood cells, skeletal myotubes and myogenic cells. Adult bone marrow progenitor cells and hematopoietic stem cells were observed to give rise to skeletal myotubes and myocytes, hepatocytes, neurons and glial cells. Adult skin stem cells have been reported to differentiate into neurons, glial cells, smooth muscle cells and adipocytes. These data show that adult-derived stem cells might not be restricted to generate tissue specific cell types, but appear to have a wider differentiation potential than previously thought. Though some of these observations have been attributed to phenomenon such as artifacts, contamination, transformation, transdifferentiation or cell fusion, certain reports suggest pluripotentiality as possible explanation for the phenotypes observed in these studies. However, the broader potential of adult stem cells remains the subject of debates and controversies, and the "true" potential of adult stem cells remains to be determined. It is hypothesized that the environment or niches in which stem cells reside may hold the key to the developmental potential of adult stem cells.

Cellular Therapy

The adult CNS is seeded with neural progenitor and stem cells. The stimulation of endogenous neural progenitor cells would represent a strategy to promote regeneration in the diseased brain and after CNS injury. Neural progenitor and stem cells have been isolated and characterized *in vitro* from various areas, neurogenic and non-neurogenic, of the adult CNS, suggesting that they may reside throughout the adult CNS. Hence, regeneration could be promoted by locally stimulating neural progenitor and stem cells at sites of degeneration. Alternatively, after experimental lesions, in the diseased brain, and after CNS injuries, such as in the brain of patients with Huntington's disease and in experimental models of strokes, new neuronal cells are generated at the sites of degeneration, such as the striatum and the cortex [95-100]. Cell tracking studies revealed that the newly generated neuronal cells originate from the SVZ, migrate partially through the RMS to the sites of degeneration, where they differentiate into the phenotype of the degenerated nerve cells [101, 102]. The identification of the SVZ as a source of newly generated neuronal cells in the diseased brain and after injury has tremendous consequence for cellular therapy in the adult CNS, as strategies to promote regeneration and repair may focus on stimulating SVZ neurogenesis.

Although new neuronal cells are generated at the sites of injuries in the brains of patients, and in animal models of neurological diseases and injuries, progressive cell loss and damages are still occurring, and no functional recovery is achieved. The generation of new neuronal cells at the site of degeneration is thus insufficient to promote functional recovery in neurological diseases and after injuries. Several hypotheses can be formulated to explain the limited regenerative capacity of the CNS. First, the number of new neurons generated is too low, albeit its stimulation to compensate for the neuronal loss -it is estimated that 0.2% of the degenerated nerve cells in the striatum after focal ischemia are replaced [101, 102]-. Second, the neurons that are produced may be non-functional because they do not develop into fully mature neurons, because they do not develop into the right type of neurons, or because they do not integrate into the surviving brain circuitry. In support of this contention, after strokes, most of the newly generated neuronal cells did not penetrate the core of the infarct where cell loss occurred, but remained in the penumbra surrounding the lesion [101, 102]. The generation of new neuronal cells at the sites of degeneration and injuries, where they replaced some of the lost nerve cells, may therefore represent a mechanism directed towards the replacement of dead or damaged neurons, and thus an attempt by the CNS to repair itself [103, 104]. The formation of glial scar tissues at the sites of injuries and degenerations is a landmark of CNS diseases and injuries, and the astrogliotic scar is felt to be an impediment to regeneration as it forms a barrier that repels growth cones [105, 106]. Therefore, the microenvironment of the diseased or injured brain may be toxic for the newly generated neuronal cells; the glial scar may limit the regeneration process [107-110]. An understanding of the molecular mechanism underlying the glial scar formation and its activity of neural progenitor and stem cells may hold the clues to the regenerative potential of adult NSCs.

Neural progenitor and stem cells can be isolated and cultured *in vitro* from the adult CNS, including from human biopsies and *post-mortem* tissues [111-113], providing a source of tissue for cellular therapy. Neural progenitor and stem cells have been grafted in various models of diseases and injuries validating their potential for the treatments of a broad range of CNS diseases [32, 114, 115]. *Ex vivo* studies revealed that grafted neural progenitor and stem cells derived from the spinal cord adopt the fate of the stem cells in the niches in which they are transplanted [32]. Hence, the microenvironment may be a determining factor for the efficiency of the recovery after transplantation, and rendering the environment favorable to the graft integration, differentiation will become a major source of investigations in the future [116, 117]. The *in vitro* differentiation of NSCs to the desired phenotype, followed by engraftment could also prove to be an alternative. However, such strategies may jeopardize the properties of neural progenitor and stem cells to integrate the host tissues.

The broader potential of adult stem cells could have tremendous potential for stem cell therapy, particularly in the CNS, as NSCs could be derived from other tissues, such as the skin [86, 87], permitting autologous therapies in which a patient's own cells are removed from the skin, grown, multiplied in a dish, and transplanted back into the patient, thereby eliminating the potential need for surgery and overcoming the issue of immune rejection or related immune-suppressing drug therapies. However, some reports have raised concerns that indeed the broader potential of adult-derived stem cells could derive from phenomena such as transformation, transdifferentiation and cell fusion [88-91], which can affect the use of adult stem cells for therapy. So the "true" potential adult stem cells remains to be fully characterized, before such strategy can be used for cell therapy.

Altogether these data show that the stimulation of endogenous neural progenitor and stem cells, and the transplantation of adult-derived neural progenitor and stem cells may represent valid strategies for the treatment of a broad range of CNS diseases and injuries. However, the microenvironment may represent a major limitation to the efficiency of the therapeutic potential of adult stem cells.

Conclusion

NSCs reside in specialized microenvironments or niches that regulate their fate. The neurogenic niches also serve as mediators between NSCs and the body, and evidences show that NSCs contribute to the formation of the neurogenic niches. Hence, NSCs and their niches are in constant, dynamic interactions with each other, regulating stem cells activity and function. In the adult brain, on the one hand, new neuronal cells are generated at sites of degeneration and injuries, but no functional recovery is achieved. On the other hand, grafted neural progenitor and stem cells adult brain integrate the host tissue, and adopt the fate as determined by the local environment. Hence, niches in the adult brain may hold the key of the developmental and therapeutic potential of adult NSCs. Further studies will aim at unraveling the molecular and cellular mechanisms underlying the interaction of adult NSCs and their microenvironments, to modify the activity of the niches, to promote the therapeutic potential of adult NSCs and to achieve functional recovery in the diseased and injured CNS.

Acknowledgments

Reproduced with kind permission of Springer Science and Business Media: Taupin P. Adult neural stem cells, neurogenic niches and cellular therapy. Stem Cell Reviews 2(3): 213-19. Copyright 2006, Springer Science and Business Media.

References

[1] Watt, F. M. and Hogan, B. L. (2000), *Science* 287, 1427-1430.

[2] Alvarez-Buylla, A. and Lim, D. A. (2004), *Neuron* 41, 683-686.

[3] Gross, C. G. (2000), *Nat. Rev. Neurosci.* 1, 67-73.

[4] Taupin, P. and Gage, F. H. (2002), *J. Neurosci. Res.* 69, 745-749.

[5] Bayer, S. A., Yackel, J. W. and Puri, P. S. (1982), *Science* 216, 890-892.

[6] Cameron, H. A., Woolley, C. S., McEwen, B. S. and Gould, E. (1993), *Neurosci.* 56; 337-344.

[7] Corotto, F .S., Henegar, J. A. and Maruniak J. A. (1993), *Neurosci. Lett.* 149, 111-114.

[8] Luskin, M. B. (1993), *Neuron* 11, 173-189.

[9] Lois, C. and Alvarez-Buylla, A. (1994), *Science* 264, 1145-1148.

[10] Lois, C., Garcia-Verdugo, J. M. and Alvarez-Buylla, A. (1996), *Science* 271, 978-981.

[11] Doetsch, F. and Alvarez-Buylla, A. (1996), *Proc. Natl. Acad. Sci. USA* 93, 14895-14900.

[12] Eriksson, P. S., Perfilieva, E., Bjork-Eriksson, T., Alborn, A. M., Nordborg, C., Peterson, D. A. and Gage, F. H. (1998), *Nat. Med.* 4, 1313-1317.

[13] Sanai, N., Tramontin, A. D., Quinones-Hinojosa, A., Barbaro, N. M., Gupta, N., Kunwar, S., Lawton, M. T., McDermott, M. W., Parsa, A. T., Manuel-Garcia Verdugo, J., Berger, M. S. and Alvarez-Buylla, A. (2004), *Nature* 427, 740-744.

[14] Gage, F. H. (2000), *Science* 287, 1433-1438.

[15] Reynolds, B. A. and Weiss, S. (1992), *Science* 255, 1707-1710.

[16] 16 Gritti, A., Parati, E. A., Cova, L., Frolichsthal, P., Galli, R., Wanke, E., Faravelli, L., Morassutti, D. J., Roisen, F., Nickel, D. D. and Vescovi, A. L. (1996), *J. Neurosci.* 16, 1091-1100.

[17] Gage, F. H., Coates, P. W., Palmer, T. D., Kuhn, H. G., Fisher, L. J., Suhonen, J. O., Peterson, D.A., Suhr, S. T. and Ray, J. (1995), *Proc. Natl. Acad. Sci. USA* 92, 11879-11883.

[18] Palmer, T. D., Takahashi, J. and Gage, F. H. (1997), *Mol. Cell. Neurosci.* 8, 389-404.

[19] Seaberg, R. M. and van der Kooy, D. (2002), *J. Neurosci.* 22, 1784-1793.

[20] Bull, N. D. and Bartlett, P. F. (2005), *J. Neurosci.* 25, 10815-10821.

[21] Taupin, P. (2006), *Ann. Acad. Med. Singapore* 35, 814-817.

[22] Johansson, C. B., Momma, S., Clarke, D. L., Risling, M., Lendahl, U. and Frisen, J. (1999), *Cell* 96, 25-34.

[23] Lendahl, U., Zimmerman, L. B. and McKay, R. D. (1990), *Cell* 60, 585-595.

[24] Morshead, C. M., Reynolds, B. A., Craig, C. G., McBurney, M. W., Staines, W. A., Morassutti, D., Weiss, S. and van der Kooy, D. (1994), *Neuron* 13, 1071-1082.

[25] Doetsch, F., Caille, I., Lim, D.A., Garcia-Verdugo, J. M. and Alvarez-Buylla, A. (1999), *Cell* 97, 703-716.
[26] Seri, B., Garcia-Verdugo, J. M., McEwen, B. S. and Alvarez-Buylla, A. (2001), *J. Neurosci.* 21, 7153-7160.
[27] Filippov, V., Kronenberg, G., Pivneva, T., Reuter, K., Steiner, B., Wang, L. P., Yamaguchi, M., Kettenmann, H. and Kempermann, G. (2003), *Mol. Cell. Neurosci.* 23, 373-382.
[28] Morshead, C. M., Garcia, A. D., Sofroniew, M. V. and van Der Kooy, D. (2003), *Eur. J. Neurosci.* 18, 76-84.
[29] Imura, T., Kornblum, H. I. and Sofroniew, M. V. (2003), *J. Neurosci.* 23, 2824-2832.
[30] Garcia, A. D., Doan, N. B., Imura, T., Bush, T. G. and Sofroniew, M. V. (2004), *Nat. Neurosci.* 7, 1233-1241.
[31] Horner, P. J., Power, A. E., Kempermann, G., Kuhn, H. G., Palmer, T. D., Winkler, J., Thal, L. J. and Gage, F. H. (2000), *J. Neurosci*. 20, 2218-2228.
[32] Shihabuddin, L. S., Horner, P. J., Ray, J. and Gage, F. H. (2000), *J. Neurosci*. 20, 8727-8735.
[33] Nishimura, E. K., Jordan, S. A., Oshima, H., Yoshida, H., Osawa, M., Moriyama. M., Jackson, I. J., Barrandon, Y., Miyachi, Y. and Nishikawa, S. (2002), *Nature* 416, 854-860.
[34] Hackney, J. A., Charbord, P., Brunk, B. P., Stoeckert, C. J., Lemischka, I. R. and Moore, K. A. (2002), *Proc. Natl. Acad. Sci. USA* 99, 13061-13066.
[35] Suda, T., Arai, F. and Hirao, A. (2005), *Trends Immunol*. 26, 426-433.
[36] Rendl, M., Lewis, L. and Fuchs, E. (2005), *PLoS Biol*. 3, e331.
[37] Martin, M. A. and Bhatia, M. (2005), *Stem Cells Dev*. 14, 493-504.
[38] Gekas, C., Dieterlen-Lievre, F., Orkin, S. H. and Mikkola, H. K. (2005), *Dev. Cell*. 8, 365-375.
[39] Wilson, A. and Trumpp, A. (2006), *Nat. Rev. Immunol*. 6, 93-106.
[40] Theise, N. D. (2006), *Am. J. Physiol. Gastrointest. Liver Physiol*. 290, G189-193.
[41] Palmer, T. D., Willhoite, A. R, and Gage, F. H. (2000), *J. Comp. Neurol*. 425, 479-494.
[42] Shen, Q., Goderie, S. K., Jin, L., Karanth, N., Sun, Y,, Abramova, N., Vincent, P., Pumiglia, K. and Temple, S. (2004), *Science* 304, 1338-1340.
[43] Lim, D. A. and Alvarez-Buylla, A. (1999), Proc. Natl. Acad. Sci. U S A 96, 7526-7531.
[44] Song, H., Stevens, C. F. and Gage, F. H. (2002), *Nature* 417, 39-44.
[45] Coultas, L., Chawengsaksophak, K. and Rossant, J. (2005), *Nature* 438, 937-945.
[46] Sun, Y., Jin, K., Childs, J. T., Xie, L., Mao, X. O. and Greenberg, D. A. (2006), *Dev. Biol*. 289, 329-335.
[47] Lyden, D., Young, A. Z., Zagzag, D., Yan, W., Gerald, W., O'Reilly, R., Bader, B. L., Hynes, R. O., Zhuang, Y., Manova, K. and Benezra, R. (1999), *Nature* 401, 670-677.
[48] Capela, A, and Temple, S. (2002), *Neuron* 35, 865-875.
[49] Conover, J. C., Doetsch, F., Garcia-Verdugo, J. M., Gale, N. W., Yancopoulos, G. D. and Alvarez-Buylla, A. (2000), *Nat. Neurosci*. 3, 1091-1097.
[50] Brantley-Sieders, D. M. and Chen, J. (2004), *Angiogenesis* 7, 17-28.
[51] Taupin, P., Ray, J., Fischer, W. H., Suhr, S. T., Hakansson, K., Grubb, A. and Gage, F. H. (2000), *Neuron* 28, 385-397.

[52] Yasuhara, O., Hanai, K., Ohkubo, I., Sasaki, M., McGeer, P. L. and Kimura, H. (1993), *Brain Res*. 628, 85-92.

[53] Gacko, M., Chyczewski, L. and Chrostek, L. (1999), *Pol. J. Pathol*. 50, 83-86.

[54] Battista, D., Ferrari, C. C., Gage, F. H. and Pitossi, F. J. (2006), *Eur. J. Neurosci*. 23, 83-93.

[55] Yue, Z., Jiang, T. X., Widelitz, R. B. and Chuong, C. M. (2005), *Nature* 438, 1026-1029.

[56] Lim, D. A., Tramontin, A. D., Trevejo, J. M., Herrera, D. G., Garcia-Verdugo, J. M. and Alvarez-Buylla, A. (2000), *Neuron* 28, 713-726.

[57] Fan, X. T., Xu, H. W., Cai, W. Q., Yang, H. and Liu, S. (2004), *Neurosci. Lett*. 366, 107-111.

[58] Lai, K., Kaspar, B. K., Gage, F. H. and Schaffer, D. V. (2003), *Nat. Neurosci*. 6, 21-27. Erratum in: (2003), Nat. Neurosci. 6, 645.

[59] Lie, D. C., Colamarino, S. A., Song, H. J., Desire, L., Mira, H., Consiglio, A., Lein, E. S., Jessberger, S., Lansford, H., Dearie, A. R. and Gage, F. H. (2005), *Nature* 437, 1370-1375.

[60] Ueki, T., Tanaka, M., Yamashita, K., Mikawa, S., Qiu, Z., Maragakis, N. J., Hevner, R. F., Miura, N., Sugimura. H. and Sato, K. (2003), *J. Neurosci*. 23, 11732-11740.

[61] Ohlstein, B., Kai, T., Decotto, E. and Spradling, A. (2004), *Curr. Opin. Cell. Biol*. 16, 693-699.

[62] Qian, X., Shen, Q., Goderie, S. K., He, W., Capela, A., Davis, A. A. and Temple, S. (2000), *Neuron* 28, 69-80.

[63] Liu, X., Wang, Q., Haydar, T. F. and Bordey, A. (2005), *Nat. Neurosci*. 8, 1179-1187.

[64] Hagg, T. (2005), *Trends Neurosci*. 28, 589-595.

[65] Ma, D. K., Ming, G. L. and Song, H. (2005), *Curr. Opin. Neurobiol*. 15, 514-520.

[66] Wurmser, A. E., Nakashima, K., Summers, R. G., Toni, N., D'Amour, K. A., Lie, D. C. and Gage, F. H. (2004), *Nature* 430, 350-356.

[67] Bjornson, C. R., Rietze, R. L., Reynolds, B. A., Magli, M. C. and Vescovi, A. L. (1999), *Science* 283, 534-537.

[68] Galli, R., Borello, U., Gritti, A., Minasi, M. G., Bjornson, C., Coletta, M., Mora, M., De Angelis, M. G., Fiocco, R., Cossu, G. and Vescovi, A. L. (2000), *Nat. Neurosci*. 3, 986-991.

[69] Clarke, D. L., Johansson, C. B., Wilbertz, J., Veress, B., Nilsson, E., Karlstrom, H., Lendahl, U. and Frisen, J. (2000), *Science* 288, 1660-1663.

[70] Rietze, R. L., Valcanis, H., Brooker, G. F., Thomas, T., Voss, A. K. and Bartlett, P. F. (2001), *Nature* 412, 736-739.

[71] Liu, Z. and Martin, L. J. (2004), *J. Neurotrauma*. 21, 1479-1499.

[72] Jackson, K. A., Mi, T. and Goodell, M. A. (1999), *Proc. Natl. Acad. Sci. USA* 96, 14482-14486.

[73] Cao, B., Zheng, B., Jankowski, R. J., Kimura, S., Ikezawa, M., Deasy, B., Cummins, J., Epperly, M., Qu-Petersen, Z. and Huard, J. (2003), *Nat. Cell. Biol*. 5, 640-646.

[74] Alessandri, G., Pagano, S., Bez, A., Benetti, A., Pozzi, S., Iannolo, G., Baronio, M., Invernici, G., Caruso, A., Muneretto, C., Bisleri, G. and Parati, E. (2004), *Lancet* 364, 1872-1883.

[75] Ferrari, G,, Cusella-De Angelis, G., Coletta, M., Paolucci, E., Stornaiuolo, A., Cossu, G. and Mavilio, F. (1998), *Science* 279, 1528-1530.

[76] Orlic, D., Kajstura, J., Chimenti, S., Jakoniuk, I., Anderson, S. M., Li, B., Pickel, J., McKay, R., Nadal-Ginard, B., Bodine, D. M., Leri, A. and Anversa, P. (2001), *Nature* 410, 701-705.

[77] Corbel S. Y., Lee, A., Yi, L., Duenas, J., Brazelton, T. R., Blau, H. M., and Rossi, F. M. (2003), *Nat. Med.* 9, 1528-1532.

[78] Petersen, B. E., Bowen, W. C., Patrene, K. D., Mars, W. M., Sullivan, A. K., Murase, N., Boggs, S. S., Greenberger, J. S. and Goff, J. P. (1999), *Science* 284, 1168-1170.

[79] Lagasse, E., Connors, H., Al-Dhalimy, M., Reitsma. M., Dohse, M., Osborne, L., Wang, X., Finegold, M., Weissman, I. L. and Grompe, M. (2000), *Nat. Med.* 6, 1229-1234.

[80] Alison, M. R., Poulsom, R., Jeffery, R., Dhillon, A. P., Quaglia, A., Jacob, J., Novelli, M., Prentice, G., Williamson, J. and Wright, N. A. (2000), *Nature* 406, 257.

[81] Eglitis, M. A. and Mezey, E. (1997), *Proc. Natl. Acad. Sci. USA* 94, 4080-4085.

[82] Brazelton, T. R., Rossi, F. M., Keshet, G. I. and Blau, H. M. (2000), *Science* 290, 1775-1779.

[83] Mezey, E., Chandross, K. J., Harta, G., Maki, R. A. and McKercher, S. R. (2000), *Science* 290, 1779-1782.

[84] Woodbury, D., Schwarz, E. J., Prockop, D. J. and Black, I. B. (2000), *J. Neurosci. Res.* 61, 364-370.

[85] Crain, B. J., Tran, S. D. and Mezey, E. (2005), *J. Neurol. Sci.* 233, 121-123.

[86] Toma, J. G., Akhavan, M., Fernandes, K. J., Barnabe-Heider, F., Sadikot. A., Kaplan, D. R. and Miller, F. D. (2001), *Nat. Cell. Biol.* 3, 778-784.

[87] Joannides, A., Gaughwin, P., Schwiening, C., Majed, H., Sterling, J., Compston, A. and Chandran, S. (2004), *Lancet.* 364, 172-178.

[88] Weissman, I. L., Anderson, D. J. and Gage, F. H. (2001), *Annu. Rev. Cell Dev. Biol.* 17, 387-403.

[89] D'Amour, K. A. and Gage, F. H. (2002), *Nat. Med.* 8, 213-214.

[90] Wurmser, A. E., Gage, F. H. (2002), *Nature* 416, 485-487.

[91] Mezey E. (2004), *Stem Cells Dev.* 13, 147-52.

[92] Qu, T. Y., Dong, X. J., Sugaya, I., Vaghani, A., Pulido, J. and Sugaya. K. (2004), *Restor. Neurol. Neurosci.* 22, 459-468.

[93] Burns, T. C., Ortiz-Gonzalez, X. R., Gutierrez-Perez, M., Keene, C. D., Sharda, R., Demorest, Z. L., Jiang, Y., Nelson-Holte, M., Soriano, M., Nakagawa, Y., Luquin, M. R., Garcia-Verdugo, J. M., Prosper, F., Low, W. C. and Verfaillie, C. M. (2006), *Stem Cells* 24, 1121-1127.

[94] Murasawa, S., Kawamoto, A., Horii, M., Nakamori, S. and Asahara, T. (2005), Arterioscler. *Thromb. Vasc. Biol.* 25, 1388-1394.

[95] Curtis MA, Penney EB, Pearson AG, van Roon-Mom WM, Butterworth NJ, Dragunow M, Connor B, Faull RL. (2003), *Proc. Natl. Acad. Sci. USA* 100, 9023-7.

[96] Tattersfield AS, Croon RJ, Liu YW, Kells AP, Faull RL, Connor B. (2004), *Neurosci.* 127, 319-32.

[97] Zhang RL, Zhang ZG, Zhang L, Chopp M. (2001), *Neurosci.* 105, 33-41.

[98] Jiang W, Gu W, Brannstrom T, Rosqvist R, Wester P. (2001), *Stroke* 32, 1201-7.

[99] Parent JM, Vexler ZS, Gong C, Derugin N, Ferriero DM. (2002), *Ann. Neurol.* 52, 802-13.

[100] Li Y, Chen J, Chopp M. (2002), *J. Neurol. Sci.* 193, 137-46.

[101] Arvidsson A, Collin T, Kirik D, Kokaia Z, Lindvall O. (2002), *Nat. Med.* 8, 963-70.

[102] Jin K, Sun Y, Xie L, Peel A, Mao XO, Batteur S, Greenberg DA. (2003), *Mol. Cell Neurosci.* 24, 171-89.

[103] Taupin, P. (2006), *Restor. Neurol. Neurosci.* 24, 9-15.

[104] Taupin, P. (2006), *Curr. Neurovasc. Res.* 3, 67-72.

[105] McGraw, J., Hiebert, G. W. and Steeves, J. D. (2001), *J. Neurosci. Res.* 63, 109-115.

[106] Ribotta, M. G., Menet, V. and Privat, A. (2004), *Acta. Neurochir. Suppl.* 89, 87-92.

[107] McIntosh, T.K., Smith, D. H., Meaney. D. F., Kotapka, M. J., Gennarelli, T. A. and Graham, D. I. (1996), *Lab. Invest.* 74, 315-342.

[108] Ghirnikar, R. S., Lee, Y. L. and Eng, L. F. (1998), *Neurochem. Res.* 23, 329-340.

[109] Stoll, G., Jander, S. and Schroeter, M. (2002), *Adv. Exp. Med. Biol.* 513, 87-113.

[110] Schwartz, M. (2003), *J. Cereb. Blood Flow Metab.* 23, 385-394.

[111] Roy, N. S., Wang, S., Jiang, L., Kang, J., Benraiss, A., Harrison-Restelli, C., Fraser, R. A., Couldwell, W. T., Kawaguchi, A., Okano, H., Nedergaard, M. and Goldman, S. A. (2000), *Nat. Med.* 6, 271-277.

[112] Palmer, T. D., Schwartz, P. H., Taupin, P., Kaspar, B., Stein, S. A. and Gage, F. H. (2001), *Nature* 411, 42-43.

[113] Schwartz, P. H,. Bryant, P. J., Fuja, T. J., Su, H., O'Dowd, D. K. and Klassen, H. (2003), *J. Neurosci. Res.* 74, 838-851.

[114] Pluchino, S., Quattrini, A., Brambilla, E., Gritti, A., Salani, G., Dina, G., Galli, R., Del Carro, U., Amadio, S., Bergami, A., Furlan, R., Comi, G., Vescovi, A. L. and Martino, G. (2003), *Nature* 422, 688-694.

[115] Cummings, B. J., Uchida, N., Tamaki, S. J., Salazar, D. L., Hooshmand, M., Summers, R., Gage, F, H. and Anderson, A. J. (2005), *Proc. Natl. Acad. Sci. USA* 102, 14069-14074.

[116] Quesenberry, P. J., Colvin, G. and Abedi, M. (2005), *Exp. Hematol.* 33, 9-19.

[117] Adams, G. B., Chabner, K. T., Alley, I. R., Olson, D. P., Szczepiorkowski, Z. M., Poznansky, M. C., Kos, C. H., Pollak, M. R., Brown, E. M. and Scadden, D. T. (2006), *Nature* 439, 599-603.

Chapter IV

Neurogenic Factors are Targets in Depression

Abstract

The confirmation that neurogenesis occurs in the adult brain and neural stem cells (NSCs) reside in the adult central nervous system (CNS) opens new avenues for our understanding of the physio-and pathology of the nervous system, as well as for therapy. Reports show that stress and antidepressants modulate neurogenesis in the adult hippocampus, and that the activity of antidepressants is mediated by adult neurogenesis. The mechanisms underlying the involvement of adult neurogenesis in depression and the activity of antide-pressants might be mediated by trophic factors and cytokines. Hence, trophic factors, cytokines and their signaling pathways are potential targets in depression, and offer new opportunities to treat this disorder.

Introduction

Contrary to a long-held dogma, neurogenesis occurs in the adult mammalian brain, including in human [1,2]. It occurs primarily in two regions of the adult brain, the dentate gyrus (DG) of the hippocampus and the subventricular (SVZ) [3]. The confirmation that neurogenesis occurs in the adult mammalian brain, and the isolation and characterization of adult-derived neural progenitor and stem cells in vitro open new opportunities for cellular therapy; the stimulation of endogenous neural progenitor or stem cells, and the transplantation of neural progenitor and stem cells, to repair the degenerated or injured pathways [4]. Adult neurogenesis is modulated by a broad range of stimuli and conditions, including environmental enrichment, physiological processes, pathological conditions, trophic factors/cytokines and drugs [5,6]. Several studies have reported that stress and antidepressants modulate neurogenesis in the adult hippocampus. Hence, adult neurogenesis could be as important for cellular therapy, as for pharmacology, of the nervous system, particularly for depression.

Adult Neurogenesis and Depression

Stress is an environmental and causal factor in precipitating episodes of depression in human [7]. Neurogenesis is decreased in the hippocampus of adult monkeys and rats subjected to psychosocial and physical stress, like the establishment of dominant/subordinate relationship between two males unknown to each other, and acute or chronic restraint [8,9]. Chronic administration of antidepressants, like the selective serotonin reuptake inhibitor (SSRI) fluoxetine and the melatonergic agonist and serotoninergic antagonist agomelatine, increases neurogenesis in the DG, but not SVZ, of adult rats and nonhuman primates [10–13]. A postmortem study performed from the brains of patients with major depression reveals that neurogenesis is not altered in the hippocampus of those patients [14].

These results suggest that neurogenesis in the adult hippocampus plays an important role in biology of depression. Particularly, stress-induced decrease of neurogenesis in the adult DG would be an important causal factor in precipitating episodes of depression. It is proposed that the waning and waxing of neurogenesis in the adult hippocampus are important factors, in the precipitation of and recovery from episodes of clinical depression, respectively [15]. The mechanisms underlying the modulation of adult neurogenesis in depression remain to be fully determined. Glucocorticoids, stress-related hormones and serotonin (5-hydroxytryptamine or 5-HT), aneurotransmitterimplicatedin the modulationofmood and anxiety-related disorders, are among the factor and molecule candidates in modulating neurogenesis during episodes of depression [16]. Other factor candidates in modulating neurogenesis during episodes of depression are substances released by the immune cells, like cytokines. Recent studies have reported that inflammatory reactions could be causal factors of neurological diseases and disorders, particularly depression [17]. Interleukin-6 (IL-6) is released by the immune cells and involved in inflammatory reactions of the nervous system. A recent study reports that IL-6 decreases neurogenesis in the adult hippocampus in rodents [18]. Hence, IL-6 is a candidate for mediating neurogenesis in episode of depression.

Adult Neurogenesis and the Pharmacology of Depression

The modulation of adult neurogenesis by antidepressants, particularly SSRIs like fluoxetine, suggests that it might contribute or mediate their activity. X-irradiation of the hippocampal region inhibits neurogenesis in the DG and prevents the behavioral effect of antidepressants, like fluoxetine, in adult mice [19]. The behavioral effect of the antidepressants in this study was assessed by the novelty-suppressed feeding test, a test used to assess chronic antidepressant efficacy, in 129SvEvTac mice. In these mice the neurogenic activity of SSRIs is mediated by 5-HT, as fluoxetine does not elicit any neurogenic and behavioral effects in 5-HT1A receptor null mice [19]. These results provide evidences and support that adult neurogenesis might mediate the behavioral effects of antidepressants.

The mechanisms underlying the activity of antidepressants on adult neurogenesis remain to be fully determined. Studies show that it might be mediated by trophic factors, particularly that brain-derived neurotrophic factor (BDNF) [20]. BDNF has an antidepressant effects; the level of expression of BDNF is increased in the brains of patients subjected to antidepressant treatments and the administration of BDNF increases adult neurogenesis in the hippocampus [21–23]. The antidepressant activity of BDNF would be mediated through the TrkB neurotrophin receptor and the mitogen-activated protein kinase signaling pathway [24,25]. More recently, the expression of the angiogenic factor, vascular endothelial growth factor (VEGF), has been reported to be up-regulated in the brain of rodents administered with antidepressants [26]. This up-regulation mediates the increase in hippocampal neurogenesis induced by the antidepressants and is mediated by the VEGF-signaling pathway; Flk-1-signaling pathway. VEGF has previously been reported to stimulate adult neurogenesis in vivo [27]. These results show that neurogenic factors are potential therapeutic targets for the treatment of depression [28].

Discussion

In all, these data show that adult neurogenesis, the hippocampus and trophic factors are targets in depression. There are, however, controversies and debates over the involvement of adult neurogenesis and the hippocampus in the biology of depression and the activity of antidepressants.

Adult Neurogenesis and Depression

Several studies have reported that antidepressants, including SSRIs like fluoxetine, produce their activity independently of adult neurogenesis. In a postmortem study, performed from the brains of patients with major depression revealing that neurogenesis is not altered in the hippocampus of those patients, most of the patients were on antidepressant medication [14]. This argues against a role of antidepressants in adult neurogenesis. The anxiolytic/antidepressant SNAP 94847 (N-[3-(1-{[4-(3,4-difluorophenoxy)-phenyl]methyl}(4-piperidyl))-4-methylphenyl]-2-methylpropanamide), an antagonist of the melanin-concentrating hormone receptor, stimulates the proliferation of progenitor cells in the DG, but its activity is unaltered in mice in which neurogenesis was suppressed by X-irradiation [29]. More recently, it was reported that fluoxetine produces its antidepressant activity independently of neurogenesis in certain strains of mice, like BALB/cJ mice [30]. In these mice, the activity of SSRIs, like fluoxetine, was reported not to be mediated by 5-HT1A receptor [30]. This shows that antidepressants could elicit their activity independently of adult neurogenesis and/or that their activity might not be mediated through adult neurogenesis. Hence, antidepressants, particularly SSRIs, like fluoxetine, might produce their activities via distinct mechanisms, some independent of adult neurogenesis.

The Hippocampus and Depression

The modulation of adult neurogenesis by antidepressants and the mediation of the behavioral activity of antidepressants by adult neurogenesis link adult neurogenesis and the hippocampus to depression. Previous studies have reported conflicting data over the involvement of the hippocampus in clinical depression. On the one hand, clinical magnetic resonance imaging and postmortem studies in depressive patients reveal that chronic stress and depression result in atrophy of the hippocampus, an atrophy reversed by antidepressant treatment [31–33]. On the other hand, other studies show that hippocampal volume remains unchanged in depressive patients [34,35]. A link between adult neurogenesis, atrophy and loss of nerve cells in the hippocampus also remains to be demonstrated. Hence, the involvement of the hippocampus and adult neurogenesis remains to be further evaluated and characterized. The hippocampus could not be primarily involved in clinical depression, as other areas of the brain could play a critical role in depression [36]. As for adult neurogenesis, it could be more a contributing factor of plas-ticity of the central nervous system (CNS) and a consequence, rather than a causative factor, in neurological diseases and disorders [37–40].

Limitations of Paradigms and Models to Study Adult Neurogenesis and Depression

Most studies conducted in animal models, rodents and nonhuman primates, used the bromodeoxyuridine (BrdU) labeling paradigm to assess neurogenesis. BrdU is a thymidine analog that incorporates DNA of dividing cells during the S-phase of the cell cycle, and is used for birthdating and monitoring cell proliferation [41]. BrdU is a toxic and mutagenic substance; it alters the cell cycle, and has transcriptional and translational effects. As a thymidine analog, it is not a marker for cell proliferation, but a marker for DNA synthesis. As such its use is subject to limitations and pitfalls [42,43]. Particularly, the blood–brain barrier is affected by drug treatments [44]. An increase in BrdU labeling in the brain could then originate from an increase in BrdU uptake rather than an increase in cell proliferation and neurogenesis, as a result of antidepressant treatment. Neuroinflammation could be a causal factor of neurological diseases and disorders, particularly depression [17] and decreases neurogenesis in the adult brain [45,46]. Hence, neuroinflammation, either as a causative factor of depression or after X-irradiation of the adult brain or hippocampal region, might affect the experimental read-out of neurogenesis in vivo. In all, data on adult neurogenesis in depression and activity of antidepressants on adult neurogenesis are difficult to interpret in light of these data. In addition, recent reports have questioned the validity of antidepressants currently prescribed, for the treatment of depression [47,48].

Conclusion

In all, adult neurogenesis, the hippocampus and trophic factors are targets in depression. However, the contribution of adult neurogenesis and the hippocampus to the biology of depression and its pharmacology remains to be elucidated and determined. The confirmation that adult neurogenesis occurs in the adult brain and neural stem cells (NSCs) reside in the adult CNS in mammals has tremendous implications for our understanding of the functioning of the nervous system and for therapy. Evidence that adult neurogenesis is involved in depression and antidepressant activity might lead to a better understanding of the etiology of depression and its pharmacology. Future directions involve the design and development of new antidepressants targeting specifically newborn neuronal cells of the adult mammalian brain, and their validation. Specifically, neurogenic factors and their signaling pathways offer new targets to develop drugs and strategies for the treatment of depression.

Acknowledgments

Reproduced from: Taupin P. Neurogenic factors are target in depression. Drug Discovery Today: Therapeutic Strategies (2008) 5(3):157-60. Copyright (2008), with permission from Elsevier.

References

[1] Eriksson, P.S. et al. (1998) Neurogenesis in the adult human hippocampus. *Nat. Med.* 4, 1313–1317

[2] Curtis, M.A. et al. (2007) Human neuroblasts migrate to the olfactory bulb via a lateral ventricular extension. *Science* 315, 1243–1249

[3] Taupin, P. et al. (2002) Adult neurogenesis and neural stem cells of the central nervous system in mammals. *J. Neurosci. Res.* 69, 745–749

[4] Taupin, P. (2006) The therapeutic potential of adult neural stem cells. *Curr. Opin. Mol. Ther.* 8, 225–231

[5] van Praag, H. et al. (2000) Neural consequences of environmental enrichment. *Nat. Rev. Neurosci.* 1, 191–198

[6] Taupin, P. (2005) Adult neurogenesis in the mammalian central nervous system: functionality and potential clinical interest. *Med. Sci. Monit.* 11, RA247–RA252

[7] Miura, H. et al. (2008) A link between stress and depression: shifts in the balance between the kynurenine and serotonin pathways of tryptophan metabolism and the etiology and pathophysiology of depression. *Stress* 11, 198–209

[8] Gould, E. et al. (1998) Proliferation of granule cell precursors in the dentate gyrus of adult monkeys is diminished by stress. *Proc. Natl. Acad. Sci. USA* 95, 3168–3171

[9] Pham, K. et al. (2003) Repeated restraint stress suppresses neurogenesis and induces biphasic PSA-NCAM expression in the adult rat dentate gyrus. *Eur. J. Neurosci.* 17, 879–886

[10] Malberg, J.E. et al. (2000) Chronic antidepressant treatment increases neurogenesis in adult rat hippocampus. *J. Neurosci.* 20, 9104–9110

[11] Malberg, J.E. et al. (2003) Cell proliferation in adult hippocampus is decreased by inescapable stress: reversal by fluoxetine treatment. *Neuropsychopharmacology* 28, 1562–1571

[12] Banasr, M. et al. (2006) Agomelatine, a new antidepressant, induces regional changes in hippocampal neurogenesis. *Biol. Psychiatry* 59, 1087–1096

[13] Perera, T.D. et al. (2007) Antidepressant-induced neurogenesis in the hippocampus of adult nonhuman primates. *J. Neurosci.* 27, 4894–4901

[14] Reif, A. et al. (2006) Neural stem cell proliferation is decreased in schizophrenia, but not in depression. *Mol. Psychiatry* 11, 514–522

[15] Jacobs, B.L. et al. (2000) Adult brain neurogenesis and psychiatry: a novel theory of depression. *Mol. Psychiatry* 5, 262–269

[16] Cameron, H.A. et al. (1994) Adult neurogenesis is regulated by adrenal steroids in the dentate gyrus. *Neuroscience* 61, 203–209

[17] Minghetti, L. (2005) Role of inflammation in neurodegenerative diseases. *Curr. Opin. Neurol.* 18, 315–321

[18] Vallieres, L. et al. (2002) Reduced hippocampal neurogenesis in adult transgenic mice with chronic astrocytic production of interleukin-6. *J. Neurosci.* 22, 486–492

[19] Santarelli, L. et al. (2003) Requirement of hippocampal neurogenesis for the behavioral effects of antidepressants. *Science* 301, 805–809

[20] Groves, J.O. (2007) Is it time to reassess the BDNF hypothesis of depression? *Mol. Psychiatry* 12, 1079–1088

[21] Siuciak, J.A. et al. (1997) Antidepressant-like effect of brain-derived neurotrophic factor (BDNF). *Pharmacol. Biochem. Behav.* 56, 131–137

[22] Chen, B. et al. (2001) Increased hippocampal BDNF immunoreactivity in subjects treated with antidepressant medication. *Biol. Psychiatry* 50, 260– 265

[23] Scharfman, H. et al. (2005) Increased neurogenesis and the ectopic granule cells after intrahippocampal BDNF infusion in adult rats. *Exp. Neurol.* 192, 348–356

[24] Saarelainen, T. et al. (2003) Activation of the TrkB neurotrophin receptor is induced by antidepressant drugs and is required for antidepressant-induced behavioral effects. *J. Neurosci.* 23, 349–357

[25] Duman, C.H. et al. (2007) A role for MAP kinase signaling in behavioral models of depression and antidepressant treatment. *Biol. Psychiatry* 61, 661–670

[26] Warner-Schmidt, J.L. et al. (2007) VEGF is an essential mediator of the neurogenic and behavioral actions of antidepressants. *Proc. Natl. Acad. Sci. USA* 104, 4647–4652

[27] Schanzer, A. et al. (2004) Direct stimulation of adult neural stem cells in vitro and neurogenesis in vivo by vascular endothelial growth factor. *Brain Pathol.* 14, 237–248

[28] Warner-Schmidt, J.L. et al. (2008) VEGF as a potential target for therapeutic intervention in depression. *Curr. Opin. Pharmacol.* 8, 14–19

[29] David, D.J. et al. (2007) Efficacy of the MCHR1 antagonist N-[3-(1-{[4-(3,4-difluorophenoxy)phenyl]methyl}(4-piperidyl))-4-methylphenyl]-2-methylpropanamide (SNAP 94847) in mouse models of anxiety and depression following acute and chronic administration is independent of hippocampal neurogenesis. *J. Pharmacol. Exp. Ther.* 321, 237–248

[30] Holick, K.A. et al. (2008) Behavioral effects of chronic fluoxetine in BALB/ cJ mice do not require adult hippocampal neurogenesis or the serotonin 1A receptor. *Neuropsychopharmacology* 33, 406–417

[31] Czeh, B. et al. (2001) Stress-induced changes in cerebral metabolites, hippocampal volume, and cell proliferation are prevented by antidepressant treatment with tianeptine. *Proc. Natl. Acad. Sci. USA* 98, 12796–12801

[32] Campbell, S. et al. (2004) Lower hippocampal volume in patients suffering from depression: a meta-analysis. *Am. J. Psychiatry* 161, 598–607

[33] Colla, M. et al. (2007) Hippocampal volume reduction and HPA-system activity in major depression. *J. Psychiatr. Res.* 41, 553–560

[34] Inagaki, M. et al. (2004) Hippocampal volume and first major depressive episode after cancer diagnosis in breast cancer survivors. *Am. J. Psychiatry* 161, 2263–2270

[35] Bielau, H. et al. (2005) Volume deficits of subcortical nuclei in mood disorders. A postmortem study. *Eur. Arch. Psychiatry Clin. Neurosci.* 255, 401–412

[36] Ebmeier, K.P. et al. (2006) Recent developments and current controversies in depression. *Lancet* 367, 153–167

[37] Taupin, P. (2006) Adult neurogenesis and neuroplasticity. *Restor. Neurol. Neurosci.* 24, 9–15

[38] Taupin, P. (2006) Neurogenesis and the effects of antidepressants. *Drug Target Insights* 1, 13–17

[39] Taupin, P. (2008) Adult neurogenesis pharmacology in neurological diseases and disorders. *Exp. Rev. Neurother.* 8, 311–320

[40] Thompson, A. et al. (2008) Changes in adult neurogenesis in neurodegenerative diseases: cause or consequence? *Genes Brain Behav.* 7, 28–42

[41] Miller, M.W. et al. (1988) Use of bromodeoxyuridine-immunohistochemistry to examine the proliferation, migration and time of origin of cells in the central nervous system. *Brain Res.* 457, 44–52

[42] Taupin, P. (2007) BrdU immunohistochemistry for studying adult neurogenesis: paradigms, pitfalls, limitations, and validation. *Brain Res. Rev.* 53, 198–214

[43] Taupin, P. (2007) Protocols for studying adult neurogenesis: insights and recent developments. *Reg. Med.* 2, 51–62

[44] Desai, B.S. et al. (2007) Blood–brain barrier pathology in Alzheimer's and Parkinson's disease: implications for drug therapy. *Cell Transplant.* 16, 285–299

[45] Ekdahl, C.T. et al. (2003) Inflammation is detrimental for neurogenesis in adult brain. *Proc. Natl. Acad. Sci. USA* 100, 13632–13637

[46] Monje, M.L. et al. (2003) Inflammatory blockade restores adult hippocampal neurogenesis. *Science* 302, 1760–1765

[47] Kirsch, I. et al. (2008) Initial severity and antidepressant benefits: a meta-analysis of data submitted to the Food and Drug Administration. *PLoS Med.* 5, e45

[48] Turner, E.H. et al. (2008) Selective publication of antidepressant trials and its influence on apparent efficacy. *N. Engl. J. Med.* 358, 252–260

Chapter V

Adult Neurogenesis, Neural Stem Cells and Alzheimer's Disease: Developments, Limitations, Problems and Promises*

Abstract

Alzheimer's disease (AD) is an irreversible progressive neurodegenerative disease, leading to severe incapacity and death. It is the most common form of dementia among older people. AD is characterized in the brain by amyloid plaques, neurofibrillary tangles, neuronal degeneration, aneuploidy and enhanced neurogenesis and by cognitive, behavioural and physical impairments. Inherited mutations in several genes and genetic, acquired and environmental risk factors have been reported as causes for developing the disease, for which there is currently no cure. Current treatments for AD involve drugs and occupational therapy, and future developments involve early diagnosis and stem cell therapy. In this manuscript, we will review and discuss the recent developments, limitations, problems and promises on AD, particularly related to aneuploidy, adult neurogenesis, neural stem cells (NSCs) and cellular therapy. Though adult neurogenesis may be beneficial for regeneration of the nervous system, it may underlie the pathogenesis of AD. Cellular therapy is a promising strategy for AD. Limitations in protocols to establish homogeneous populations of neural progenitor and stem cells and niches for neurogenesis need.

Introduction

AD was first described by Alois Alzheimer in 1906. Alois Alzheimer reported the presence of amyloid or senile plaques and neurofibrillary tangles in the brain of patients with severe dementia [1]. AD is a neurodegenerative disease, associated with the loss of nerve cells in areas of the brain that are vital to memory and other mental abilities, like the entorhinal cortex, hippocampus and neocortex. The disease is characterized by progressive cognitive, behavioural and physical impairments [2, 3]. It is the most common form of dementia among elderly, with 50 to 70% of clinical cases confirmed as AD, post-mortem. Aging is the major contributing factor for increased risk of developing AD. The risk of developing AD doubles every 5 years after the age of 65. AD affects 30% of individuals of age over 80 [4]. It affects more than 26 millions of patients worldwide; this number is expected to quadruple by 2050 as population age [5].

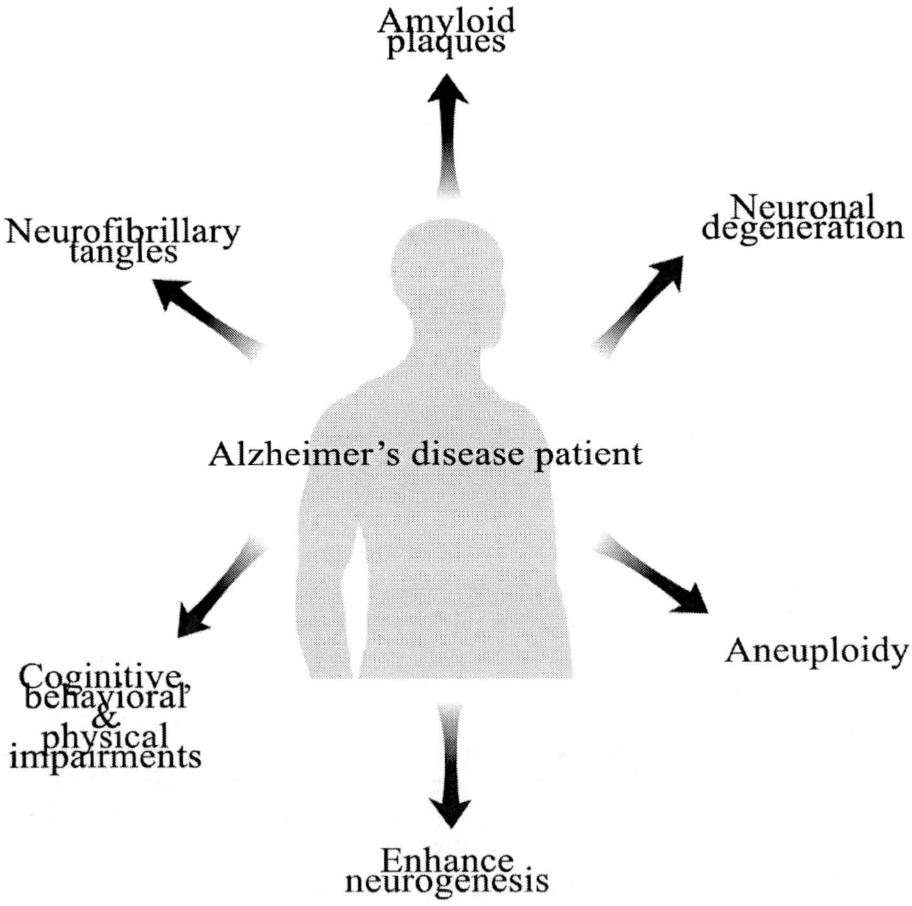

Figure 1. Alzheimer's disease is the most common form of dementia among elderly. AD is characterized in the brain by amyloid plaques, neurofibrillary tangles, neuronal degeneration, aneuploidy, enhanced neurogenesis and cognitive, behavioural and physical impairments. The origin and contribution of these processes to the etiology and pathogenesis of AD remain mostly unknown.

Neurogenesis occurs throughout adulthood in mammals [6, 7]. It occurs primarily in two regions of the adult brain, the subventricular zone (SVZ) and the dentate gyrus (DG) of the hippocampus, in various species including humans [8-10]. In the DG, newly generated neuronal cells in the sub-granular zone (SGZ) migrate to the granule cell layer, where they differentiate into granule-like cells and extend axonal projections to the CA3 region of the

Ammon's horn. Newly generated neuronal cells in the anterior part of the SVZ migrate through the rostro-migratory stream to the olfactory bulb, where they differentiate into interneurons, granule and periglomerular neurons [11]. Newly generated neuronal cells in the adult brain originate from residual stem cells [12]. NSCs are the self-renewing multipotent cells that generate the main phenotypes of the nervous system. Newly generated neuronal cells of the adult brain may be involved in a broad range of physio- and pathological processes, like learning and memory, AD, epilepsy and schizophrenia [13-16]. The confirmation that adult neurogenesis occurs in the adult brain and NSCs reside in the adult central nervous system (CNS) has tremendous implications for our understanding of development, physio- and pathology and for cellular therapy. On the one hand, newly generated neuronal cells of the adult brain would contribute to plasticity and regeneration of the nervous system [17]. The adult CNS has the potential to self-repair. On the other hand, they may be involved in the pathology and pharmacology of neurological diseases and disorders, and particularly AD [18].

In this manuscript, we will review and discuss the potential contribution of newly generated neuronal cells of the adult brain to the pathogenesis of AD and of adult NSCs for cellular therapy, for the treatment of AD.

Alzheimer's Disease Pathology and Pathogenesis

Clinical and Histopathological Diagnosis

Doctors diagnose AD primarily by symptoms, cognitive impairments, behavioural changes and risk factor assessments [19-21]. There are two forms of the disease. Late-onset AD (LOAD) refers to cases of AD diagnosed after the age of 65. Early-onset AD (EOAD) refers to cases of AD diagnosed at younger age. Most cases of LOAD are sporadic. They are believed to be caused by a combination of genetic risk factors, like the presence ApoE varepsilon 4 allele (ApoE4), acquired risk factors, like hypertension and diabetes, and environmental risk factors, like neuroinflammation and oxidative stress [22]. LOAD accounts for the majority, over 93%, of all cases of AD. In contrast, EOAD is a rare form of the disease and is mostly inherited. Inherited form of AD is also known as familial Alzheimer's disease (FAD). It is caused by mutations in so-called familial Alzheimer genes, like the gene of β-amyloid precursor protein (APP). About 200 families in the world carry the gene mutations that cause EOAD. Once diagnosed, the average life expectancy of patients with AD is 8.5 years, though the disease can last for as many as 20 years.

Amyloid plaques and neurofibrillary tangles are hallmarks of AD, Figure 1. They are deposits of proteins distributed throughout the brain of patients with AD, particularly in the entorhinal cortex, hippocampus, temporal, frontal and inferior parietal lobes. Their density increases as the disease progresses. However, the correlation between the density of amyloid plaques and the severity of the dementia is not clearly established [23]. Amyloid plaques are thought to be the first histological changes that occur in AD [24]. AD is associated initially with the loss of nerve cells in areas of the brain that are vital to memory and other mental abilities, like the entorhinal cortex, hippocampus and neocortex, but also in regions of the brain important to sense of smell. As the disease progresses, other regions of the brain are affected, including the medial temporal area, lateral hemisphere, basal forebrain and locus coeruleus. This leads progressively to the overall shrinkage of the brain, severe incapacity and death [25]. Autopsies are performed to assess the presence of amyloid plaques and neurofibrillary tangles, the extent of the degeneration and to confirm earlier diagnosis.

Amyloid Plaques and Neurofibrillary Tangles

Amyloid plaques are extracellular deposits of proteins, surrounded by degenerating nerve cells in the brain of patients with AD and in the retina of patients with aged-related macular degeneration [26]. They are composed of amyloid fibrils and α1-antichymotrypsin (ACT), a serine protease inhibitor. Protein β-amyloid is a 42 amino acid β-peptide originating from the post-transcriptional maturation of APP [27]. It is synthesized and secreted by nerve cells, as a soluble peptide. The gene for APP is located on chromosome 21q21 [28, 29]. Protein β-amyloid is an amyloidogenic protein; proteins forming amyloid fibrils [30]. Amyloidogenic proteins are monomer and soluble in their physiological state. Under pathological conditions, they form insoluble extracellular aggregates or deposits of amyloid fibrils [31]. Deposit of amyloid fibrils in the brain of patients with AD results from aggregation of protein β-amyloid. They arise when protein β-amyloid is induced to form filaments by amyloid-promoting factors expressed in certain regions of the brain or under certain gene mutations, including in *APP*. The aggregation of protein amyloid results from the abnormal processing of APP. According to the amyloid hypothesis, protein β-amyloid deposit may cause AD. As the amyloid deposits develop in the brain, the brain cells start dying and the signs and symptoms of AD begin [32, 33]. Alternatively, the over-expression of protein β-amyloid or of mutated form of the protein may be the cause leading to the pathology of AD [34]. In support of this contention, the correlation between the density of amyloid plaques and the severity of dementia is not clearly established [23]. In this latter model, the deposit of protein β-amyloid would be a consequence rather than a cause of AD. The contribution of protein β-amyloid to the etiology and pathogenesis of AD remains to be fully understood and determined.

Neurofibrillary tangles are deposits of proteins present inside neuronal cells in the brain of patients with AD. They are composed of hyperphosphorylated Tau proteins [35]. Tau protein is a microtubule-associated phosphoprotein. It is an axonal protein involved in the formation of microtubules [36]. Tau proteins interact with tubulin to form the microtubules [37]. Microtubules are involved in the structure, transport and division of cells. The TAU gene is located on chromosome 17q21.1 [38]. The phosphorylation of Tau is modulated by phosphatases and kinases. It regulates the binding of Tau to microtubules. The phosphorylation of Tau decreases the binding of Tau to microtubules. It results in instability of the microtubules and aggregation of Tau proteins. In AD and tauopathies, Tau protein is hyperphosphorylated by kinases [39]. It leads to the dissociation between Tau and tubulin. This triggers the breakdown of microtubules and the polymerization and aggregation of Tau proteins [40]. It results in the formation of neurofibrillary tangles and cell death [41].

These mechanisms underlying the formation of amyloid plaques and neurofibrillary tangles are not fully understood. They are likely to be different depending on whether AD is caused by genetic mutations or genetic, acquired or environ-mental risk factors.

Genetic Factors

EOAD is a rare form of AD, mostly inherited, caused by mutations in familial Alzheimer genes [42]. Mutations in these genes almost always result in the individual developing the disease [43]. The patients generally have a family history with EOAD. Three genes have been identified for the FAD. These are the *APP* gene, the presenilin-1 (*PSEN-1*) gene and the presenilin-2 (*PSEN-2*) gene [44]. In contrast, no single causal genetic mutation has been identified for LOAD [45]. LOAD is believed to be caused by a combination of genetic, acquired and environmental risks factors [46-50]. These risk factors increase the probability of developing the disease. Their absence does not mean that AD will not develop. The identification of gene mutations linked to LOAD has mainly been performed by single-nucleotide polymorphism studies to link the genetic polymorphism to the disease. The apolipoprotein E (APOE) gene is the best established genetic risk factor for LOAD. It accounts for the vast majority of causes and risks to develop AD [51].

Familial Alzheimer genes. APP is a 695-770 amino acid protein. It is normally synthesized and present in the brain and other tissues. APP plays a role in various cell functions, like cell adhesion and neurite growth. APP is processed by α-, β- and γ-secretase enzymes. In physiological conditions, it is cleaved by the α– and γ-secretase enzymes into a 40 amino acid β-peptide. Mutations in the APP gene cause excessive cleavage of APP, by the β- and γ-secretase enzymes. This results in increased production of the 42 amino acid β-amyloid peptide; this latter form of protein β-amyloid aggregates into insoluble amyloid deposits, particularly in the brain. The presenilin proteins are components of the γ-secretase complex. These enzymes play a role in the maturation of APP into the 42 protein β-amyloid [52]. Mutations in the *PSEN-1* gene and *PSEN-2* gene lead to excessive cleavage by γ-secretase enzyme, resulting in increased production and aggregation of protein β-amyloid [53]. The *PSEN-1* gene and the *PSEN-2* gene are located on chromosome 14q24.3 and 1q31–q42, respectively [52]. Among the cases of EOAD for which a genetic mutation has been identified as cause of the disease, 30-70% of the mutations are in the *PSEN-1* gene, 10-15% in the *APP* gene and less than 5% in the *PSEN-2* gene [21]. For many individuals/families with cases of EOAD, the genetic mutation causative of the disease has not been determined. There are other gene mutations involved in EOAD to be identified.

Genetic risk factors. Several genes have been identified as risk factors for LOAD. Among them, the *ApoE* gene and the neuronal sortilin-related receptor (*SORL1*) gene [54, 55]. ApoE is a plasma protein. It plays a role in the transport and metabolism of lipids. ApoE is a ligand for the low density lipo-protein receptors. Through interaction with these receptors, it participates in the transport of cholesterol and other lipids, to various cells of the body [56]. There are three major isoforms of ApoE, ApoE2, ApoE3 and ApoE4, encoded by different alleles in humans. The ApoE2 isoform occurs in 10%, ApoE3 in 74% and ApoE4 in 16% of white populations. Individuals who have the *ApoE4* allele have increased risk of developing AD. Up to 50% of people who have AD have at least one *ApoE4* allele. People who have two *ApoE4* alleles have a higher risk of developing AD, after age of 65 [57]. It is estimated that one copy of the *ApoE4* allele reduces the age of onset by 7-9 years [58]. The role of ApoE4 in the etiology and pathogenesis of AD remains to be established. In patients with AD, ApoE is localized in amyloid plaques and neurofibrillary tangles. ApoE4 may promote the formation of amyloid plaques, by a mechanism yet to be determined [59]. The *ApoE* gene is located on chromosome 19q13.2. SORL1 belongs to a family of proteins termed retromer. Retromers are involved in intracellular trafficking. SORL1 is involved in the trafficking and recycling of APP [60]. Reduced expression of the *SORL1* gene is associated with an increase in the risk for LOAD. It is also associated with an increase in density of amyloid plaques. The decreased expression of SORL1 is linked to variants in at least two different clusters of intronic sequences in the *SORL1* gene. The variants of *SORL1* may promote AD by suppressing the activity of the gene. This may affect the processing of APP and increase its production [55]. Variants for the genes coding for α2-macroglobulin, monoamine oxidase A and myeloperoxidase have been linked with the occurrence of LOAD. α2-Macroglobulin is a protease inhibitor found in neuritic plaques. The α2-macroglobulin gene is located on chromosome 12p13.3. Monoamine oxidase A is a regulator of the metabolism of neuroactive and vasoactive amines within the CNS. Myeloperoxidase is an enzyme present in circulating monocytes and neutrophils; it catalyses the production of the oxidant hypochlorous acid. Myeloperoxidase is thought to contribute to the pathology of AD through oxidation of either protein β-amyloid or ApoE. Studies have reported a linkage between the polymorphism of the gene for cystatin C (*CST3*) with LOAD [61]. There are also evidences that polymorphisms within the genes of the folate methionine and homocysteine metabolic pathways are involved in the pathogenesis of AD [45, 62, 63].

Other genes have been identified as promoting the risk factors of identified genetic risk factors. GRB-associated binding protein 2 (GAB2) belongs to a family of proteins that plays a central role in signalling by receptor protein-tyrosine kinases [64]. Mutations in the *GAB2* gene are linked with increasing risk of LOAD, in people with ApoE4 allele [65]. *GAB2* gene may offset some of the ApoE4 associated risks for developing AD, by inhibiting the formation of amyloid plaques, whereas mutation in the *GAB2* gene would promote the formation of amyloid plaques, in people with *ApoE4* allele.

Sporadic forms of AD are the most common cases of AD. They most generally develop after age 65. They correspond to most cases of LOAD. The genetic risk factors in sporadic forms of AD present an unclear mode of inheritance. Sporadic cases of EOAD can occur, with no family history and no identified causal genetic mutations. Cases of FAD can occur after age of 65 [66]. The causal mutations involved in these forms of LOAD remain unidentified.

Chromosome 21 and Aneuploidy

Several studies reveal that cells from patients with AD elicit aneuploidy, particularly for chromosome 21. Preparations of lymphocytes of patients with sporadic form of AD elicit an elevation in aneuploidy for chromosomes 13 and 21, particularly for chromosome 21 [67,

68]. Preparations of lymphocytes of patients with AD, familial and sporadic forms, elicit a 2-fold increase in the incidence of aneuploidy for chromosomes 18 and 21 [69]. Deposit of protein amyloid is one of the histopathological features of AD and one of the probable cause for the pathogenesis of AD. The APP gene is located on chromosome 21 [28, 29]. Aneuploidy for chromosome 21 has been proposed as one of the mechanisms underlying the pathogenesis of AD [70]. The synthesis and deposit of protein amyloid could have for origin the over-expression of mutant or wild type amyloid protein in aneuploid cells, due to the duplication of the *APP* gene that resides on chromosome 21, in patients with FAD or sporadic form of AD respectively.

Patients with Down's syndrome develop, during their 30s and 40s, dementia and neuropathology that share characteristics with AD [71-73]. Down's syndrome has for pathogenic cause trisomy for the chromosome 21 [74]. Aneuploidy for chromosome 21 would underlie the pathogenesis of the dementia that occurs in Down's syndrome and AD patients [70]. Cells that are the most likely to develop aneuploidy are dividing cells. Aneuploidy results from the non-disjunction of chromosomes during mitosis or meiosis [75]. A wide range of cells elicit aneuploidy in patients with AD [67-69]. The non-disjunction of chromosomes, particularly of chromosome 21, in stem cells and/or populations of somatic cells that retain their ability to divide could be at the origin of aneuploidy in patients with AD. The origin of aneuploidy in patients with Down's syndrome would result from the non-disjunction for chromosome 21 in germ cells, during meiosis [70].

According to this model, genetic, acquired and environmental factors that promote or contribute to aneuploidy, particularly for chromosome 21, would increase the risk of developing AD. Mutated forms of PSEN-1 are detected in the centrosomes and interphase kinetochores of dividing cells. Mutated PSEN-1 may then be involved in the segregation and migration of chromosomes [76]. Mutation in *PSEN-1* is a causative factor for EOAD. Mutated PSEN-1 may contribute to the pathogenesis of FAD not only by abnormally processing APP, but also by promoting the non-disjunction of chromosomes and aneuploidy in cells. In AD and tauopathies, Tau is hyperphosphorylated by kinases, leading to the dissociation between Tau and tubulin and the breakdown of microtubules [39, 40]. The breakdown of microtubules, by hyperphosphorylated Tau, could promote aneuploidy by causing defects in the mitotic spindle during mitosis. Hyperphosphorylated Tau is a component of neurofibrillary tangles, a histopathological hallmark of AD and a probable cause for cell death in AD patients. Hyperphosphorylated Tau protein may contribute to the pathogenesis of AD not only by the polymerization and aggregation of Tau proteins, resulting in the formation of neurofibrillary tangles and cell death [41], but also by promoting the non-disjunction of chromosomes and aneuploidy in cells. Hence, mutated PSEN-1 and hyperphosphorylation of Tau could promote aneuploidy in somatic cells, particularly for chromosome 21, leading to AD.

The *PSEN-1* and *TAU* genes are located on chromosomes 14 and 17, respectively [38, 52]. Aneuploidy for chromosomes 14 and 17 could lead to an over-expression of mutated PSEN-1 and Tau, respectively, further increasing the risk of aneuploidy and of the formation of neurofibrillary tangles. Aneuploidy for chromosomes 14 and 17 may therefore also contribute to increase the risk of developing AD and the progression of the disease. Oxidative stress promotes aneuploidy for chromosome 17 [77]. Oxidative stress is an environmental risk factor for developing AD [22]. It may act as a risk factor for AD, by promoting the expression of Tau proteins, which hyperphosphorylation causes the formation of neurofibrillary tangles. β-Amyloid, reactive oxygen species and oxidative stress induce cell cycle re-entry and neuronal death [78-81]. A "two-hit hypothesis" has been proposed to conciliate the activity of oxidative stress and abnormal mitotic signalling, like abortive cell cycle re-entry or gene duplication without cell division leading to cell death, as causative factors of AD. Oxidative stress and abnormal mitotic signalling can act independently as initiators; however both processes are necessary to propagate the pathogenesis of AD [82]. Abnormal mitotic signalling may lead to a small population of aneuploid cells that over-express genes that contribute to the development of the disease, like *APP* and *TAU*. As these cells undergo cell death, they trigger an inflammatory reaction in the regions of amyloid and neuritic plaques formation. This further promotes the development of the disease. In this model, individuals may eventually develop the disease, over a longer period of time.

In all, chromosomes non-disjunction and aneuploidy are contributing factors for the pathogenesis of AD by promoting the expression of genes involved in AD; this primarily by promoting the formation of amyloid deposits and neurofibrillary tangles.

Enhanced Neurogenesis in the Brain of Patients with Alzheimer's Disease

The expression of markers of immature neuronal cells, like doublecortin and polysialylated nerve cell adhesion molecule, is increased in hippocampal regions, particularly the DG, in the brain of patients with AD [83]. In animal models of AD, neurogenesis is enhanced in the DG of transgenic mice that express the Swedish and Indiana APP mutations, a mutant form of human APP [84]. It is decreased in the DG and SVZ of mice deficient for PSEN-1 and/or APP, in transgenic mice over expressing variants of APP or PSEN-1 [85-89]. It is decreased in the DG of PDAPP transgenic mouse, a mouse model of AD with age-dependent accumulation of protein β-amyloid [90]. Transgenic mice that express the Swedish and Indiana APP mutations, mice deficient for PSEN-1 and/or APP and transgenic mice over expressing variants of APP or PSEN-1 are transgenic mice that express variants of FAD genes. The discrepancies of the data observed on adult neurogenesis in autopsies and animal models of AD may originate from the validity of the animal models, particularly transgenic mice, as representative of AD and to study adult phenotypes [91]. Mice deficient for APP and PSEN-1 provide information on the activities and functions of the proteins involved in AD. They do not represent the disease. The effects of genetic mutations during development may have adverse effects on adult phenotypes, like adult neurogenesis. Aggregation of protein β-amyloid affects adult neurogenesis and may be an underlying of the modulation of neurogenesis in AD brain and animal models of AD [92]. These results indicate that neurogenesis is enhanced in the brain of patients with AD. It would result from damaged or stimulation induction of neurogenesis. It may be a consequence, rather than a cause, of the disease [18]. Enhanced neurogenesis in AD may contribute to a regenerative attempt, to compensate for the neuronal loss [84, 93].

Early Diagnosis and Treatments

Since FAD and sporadic forms of AD have a genetic component, it is possible to detect causative mutations and genetic risk factors for developing AD in patients, by genetic testing. A broad range of tests are being developed and validated to improve the diagnosis of AD, including the measurement of amyloid deposits by brain imaging and the proteomic analysis of cerebrospinal fluid [94-99]. These tests aim to improve the diagnosis of AD and detect AD, or the susceptibility to AD, at early stages. Early diagnosis of AD will allow providing the patients with better treatment, assistance and care [100]. AD patients could be treated earlier, in the aim to curb the progression of the disease, by identifying who is at risk and prescribing drugs and lifestyle changes to keep them healthy. However, until such knowledge is available, the use of these tests is not without ethical and moral issues for the physicians and patients [21].

There is currently no cure for AD. Actual treatments consist in drug and occupational therapies [101]. Three types of drugs are currently used to treat AD: i) blockers of the formation of amyloid deposits, like alzhemed, ii) inhibitors of acetylcholine esterase, like tacrine, galantamine and rivastigmine, and iii) N-methyl-D-aspartate glutamate receptor antagonists, like memantine [102-106]. Acetylcholine esterase inhibitors are thought to improve cognitive functions by enhancing cholinergic neurotransmission, that are affected in brain regions of AD and that are important for learning and memory. N-methyl-D-aspartate glutamate receptor antagonists confer protection against excitotoxic neurodegeneration. These drugs produce improvements in cognitive and behavioural symptoms of AD. Other treatments that are considered involve secretase inhibitors, drugs for lowering cholesterol levels, chelators of metals, anti-inflammatory drugs and protein β-amyloid vaccination, to stimulate the immune system to clean up the amyloid [107-109].

Discussion

AD is the most common form of dementia among elderly. Amyloid plaques, neurofibrillary tangles, neurodegeneration, aneuploidy and enhanced neurogenesis are landmarks of AD pathology, Figure 1. The origin and contribution of these processes to the etiology and pathogenesis of AD remain mostly unknown. Among the causes of AD are genetic mutations and genetic, acquired and environmental risk factors, neuroinflammation and oxidative stress. The confirmation that adult neurogenesis occurs in the adult brain and NSCs reside in the adult CNS not only brings new opportunities for the treatment of AD, but also raises the question of the involvement of newly generated neuronal cells of the adult brain in the etiology and pathogenesis of AD.

Aneuploid Cells in Regions of Degeneration in the Brain of Patients with Alzheimer's Disease

The adult brain contains a substantial number of cells that are aneuploids; 5 to 7% of the cells in the brain of adult mice [110, 111]. Aneuploidy may originate from non-disjunction of chromosomes during cell division, abortive cell cycle re-entry, cells undergoing DNA duplication without cell division and cell fusion [75, 112]. In the brain of patients with AD, 4 to 10% of neurons in regions of degeneration, like the hippocampus, express proteins of the cell cycle and some at-risk neurons are aneuploids [113, 114]. The marker of the phase G2 of the cell cycle, cyclin B, is expressed in neurons in regions of degeneration, particularly the hippocampus, in patients with AD [115]. Nerve cells are post-mitotic cells in the adult brain. Hence, the characterization of cyclin B and aneuploidy in neurons suggests that cells entered the cell cycle and underwent DNA replication, but did not complete the cell cycle, in regions of degeneration in the brain of AD patients. It is proposed that the genetic imbalance in aneuploid cells signifies that they are fated to die [116]. Their relatively high percentage at any one time in regions of degeneration in AD brains suggests that they will undergo a slow death process. Unlike apoptosis, these cells may live in this state for months, possibly up to 1 year [117, 118]. The deregulation and/or re-expression of proteins controlling the cell cycle of nerve cells, triggering cycle re-entry with blockage in phase G2, would underlie the neurodegenerative process and pathogenesis of AD.

Aneuploidy for chromosome 21 has been proposed as one of the contributing factors for the pathogenesis of AD [70]. APP is located on chromosome 21 and over-expression of APP would promote the formation of amyloid plaques. According to the "amyloid hypothesis", this would underlie cell death and the pathogenesis of AD. Hence, aneuploidy for chromosome 21 in neurons in regions of degeneration would underlie the pathogenesis of AD in two ways, by promoting the process cell death and neurodegeneration and the formation of amyloid plaques. The *TAU* gene is located on chromosome 17. Aneuploidy for chromosome 17 in neurons in regions of degeneration would underlie the pathogenesis of AD, by promoting the process neurodegeneration and the formation of neurofibrillary tangles.

Abortive Versus Beneficial Neurogenesis in the Adult Brain

Cells that are the most likely to develop aneuploidy are dividing cells [75]. Aneuploidy for chromosomes 21, 14 and/or 17 is a contributing factor to the pathogenesis of AD, by increasing the risk of amyloid plaques formation, aneuploidy and neurofibrillary tangles formation. It has been proposed that the non-disjunction of chromosomes, particularly of chromosomes 21, 14 and 17, in stem cells and/or populations of somatic cells that retain their ability to divide is at the origin of aneuploidy in patients with AD [70]. In the adult brain, neurogenesis occurs primarily in the SGZ and SVZ. Newly generated neuronal cells of the adult brain would originate from stem cells. Newly generated neuronal cells of the adult brain would contribute to plasticity and regeneration of the nervous system. The process of adult neurogenesis holds the potential to generate populations of aneuploid cells particularly in the neurogenic areas. The non-disjunction of chromosomes during the process of cell division of newly generated progenitor cells of the adult brain could lead to newly generated neuronal cells that are aneuploids or to a population of aneuploid cells that would not proceed with its developmental program, Figure 2. Such aneuploidy, particularly for chromosomes 21, 14 and/or 17 and particularly in the hippocampus, would contribute to the pathogenesis of AD. Cell death is a normally occurring process in the adult brain especially in the neurogenic zones, as a significant proportion of newly generated cells in the SVZ and SGZ are believed to undergo apoptosis rather than achieving maturity [119, 120]. The number of newborn neuronal cells generated in the adult brain is relatively low, particularly in the DG. It is estimated that 0.1% of the granule cell population is generated per day in the DG of young adult rodents [120, 121]. Hence, aneuploidy in newly generated neuronal cells would be a rare event. It would most likely contribute to the pathogenesis of AD, in individuals predisposed to develop the disease. This suggests that newly generated neuronal cells of the adult brain and adult neurogenesis could be involved in the pathogenesis of AD, and not only in a regenerative process.

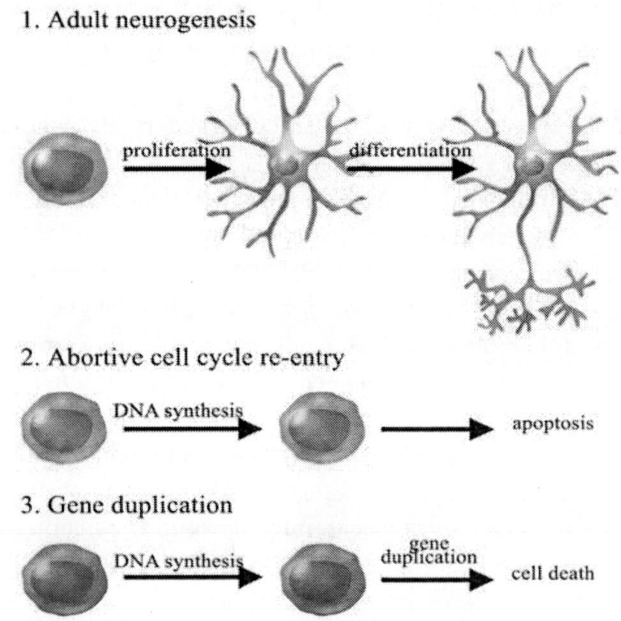

Figure 2. Abortive versus beneficial neurogenesis in the adult mammalian brain. In the adult brain, neurogenesis occurs primarily in the SGZ and SVZ. Cell death is a normally occurring process in the neurogenic zones, as a significant proportion of newly generated cells are believed to undergo apoptosis rather than achieving maturity (A). Newly generated neuronal cells of the adult brain would originate from stem cells (B). Newly generated neuronal cells of the adult brain would contribute to plasticity and regeneration of the nervous system. The process of adult neurogenesis holds the potential to generate populations of aneuploid cells, particularly in the neurogenic areas. The non-disjunction of chromosomes during the process of cell division of newly generated progenitor cells of the adult brain could lead to newly generated neuronal cells that are aneuploids (C) or to a population of aneuploid cells that would not proceed with its developmental program (D). The genetic imbalance in aneuploid cells signifies that they are fated to die. Aneuploidy in newly generated progenitor cells of the adult hippocampus, particularly for chromosomes 21, 14 and/or 17, would contribute to the pathogenesis of AD. Newly generated neuronal cells of the adult brain and adult neurogenesis could be involved in the pathogenesis of AD, and not only in a regenerative process.

Abortive Cell Cycle Re-entry and Cells Undergoing DNA Duplication without Cell Division Versus Neurogenesis in Animal Models of Alzheimer's Disease

Most studies conducted in animal models of neurological diseases and disorders, and particularly in animal models of AD, use bromodeoxyuridine (BrdU) labelling, as a paradigm to study adult neurogenesis. BrdU is a thymidine analog used for birth dating and monitoring cell proliferation [122, 123]. There are pitfalls and limitations over the use of thymidine analogs, and particularly BrdU, for studying neurogenesis [124-126]. BrdU is a toxic and mutagenic substance. It triggers cell death, the formation of teratomes, alters DNA stability, lengthens the cell cycle and has mitogenic, transcriptional and translational effects on cells that incorporate it. All of which have profound consequences on neurogenesis. In addition, as a thymidine analog, BrdU is not a marker for cell proliferation, but a marker for DNA synthesis. Therefore, studying neurogenesis with BrdU requires distinguishing cell proliferation and neurogenesis from other events involving DNA synthesis, like DNA repair, abortive cell cycle re-entry and gene duplication without cell division, leading to aneuploidy [127, 128]. In addition, despite earlier reports [129], the permeability of the blood-brain barrier may be affected in AD [130]. In these conditions, an increase in BrdU-labelling in the brain could originate from an increase in BrdU uptake rather than an increase in cell proliferation and neurogenesis [127, 128]. Cell cycle proteins, like cyclin B the marker of the phase G2, are expressed in neurons, in regions in which degeneration occurs, and some at-risk neurons in regions of degeneration are aneuploids in the brain of AD patients [113, 114]. The evidence that cell cycle re-entry and DNA duplication, without cell division, precedes neuronal death in degenerating regions of the CNS suggests that when using immunohistochemistry for proteins of the cell cycle to study adult neurogenesis, this paradigm does not allow discriminate between cells undergoing DNA duplication, without cell division, as part of their pathological fate and newly generated neuronal cells [127, 128]. Hence, data involving the use of BrdU-labelling and immunohistochemistry for proteins of the cell cycle, as paradigms for studying adult neurogenesis in neurological diseases and disorders, and particularly in AD, must be carefully assessed and analyzed.

Potential and Limitations of Adult Neurogenesis and Neural Stem Cells for the Treatment of Alzheimer's Disease

The confirmation that neurogenesis occurs in the adult brain and NSCs reside in the adult CNS, opens new opportunities for cellular therapy for a broad range of neurological diseases, disorders and injuries, particularly for neurodegenerative diseases like AD [18]. The adult CNS may be amenable to repair. To this aim, two strategies are being considered: the stimulation of endogenous neural progenitor or stem cells of the adult brain, and the transplantation of adult-derived neural progenitor and stem cells, to repair the degenerated or injured pathways [131]. There are limitations to the potential of adult NSCs for therapy. On the one hand, stem cells reside in specialized microenvironments or "niches", particularly in the adult brain [132, 133]. An angiogenic niche and an astroglial niche for neurogenesis have been identified and characterized in the adult brain. These niches regulate and control the self-renewal and differentiation activities of NSCs. The microenvironment plays therefore a key role in the therapeutic potential of adult stem cells, whether endogenous or transplanted. Unravelling and unlocking the mechanisms underlying the neurogenic niches for neurogenesis will contribute to the realization of the therapeutic potential of adult NSCs [134]. On the other hand, protocols currently established to isolate and culture neural progenitor and stem cells from the adult brain yield to heterogeneous populations of neural progenitor and stem cells, limiting their therapeutic potential [135]. Identifying markers of neural stem/progenitor cells and conditions to maintain such culture homogeneous will enhance the therapeutic potential of adult-derived neural progenitor and stem cells.

Cell grafting targets local areas of the brain. The intra-cerebral transplantation of adult-derived neural progenitor and stem cells may not be applicable for the treatment of AD, where the degeneration is widespread. Neural progenitor and stem cells, administered intravenously, migrate to diseased and injured sites of the brain [136, 137]. Systemic injection provides a non-invasive strategy for delivering neural progenitor and stem cells in the adult CNS. Experimental studies reveal that systemic injection of neural progenitor and stem cells promote functional recovery in an animal model of multiple sclerosis [137]. This shows that systemic injection provides a model of choice for delivering NSCs for the treatment of neurological diseases and injuries and may provide a paradigm of choice for the treatment of AD. Adult neurogenesis is modulated by a broad range of environmental and physio- and pathological stimuli and processes, as well as trophic factors/cytokines and drugs [138]. Conditions that stimulate endogenous neurogenesis in the adult brain may be applied to promote the regenerative and recovery processes.

Conclusion and Perspectives

AD is characterized in the brain by amyloid plaques, neurofibrillary tangles, neurodegeneration, aneuploidy and enhanced neurogenesis. The role and contribution of these processes to the etiology and pathogenesis of AD remain to be elucidated and fully understood. The confirmation that neurogenesis occurs in the adult brain and NSCs reside in the adult CNS opens new perspectives and opportunities for the treatment and cure, but also for our understanding of the etiology and pathogenesis of AD. Chromosomes non-disjunction and aneuploidy are contributing factors for the pathogenesis of AD. The non-disjunction of chromosomes during the process of cell division of newly generated progenitor cells of the adult brain could lead to newly generated neuronal cells that are aneuploids or to a population of aneuploid cells that would not proceed with its developmental program. Hence, newly generated neuronal cells of the adult brain would not only contribute to plasticity and regeneration of the nervous system, but also to the pathogenesis of neurological diseases and disorders, particularly AD. Future studies will aim at understanding the role and contribution of adult neurogenesis to the pathology of AD and to design protocols and strategies to treat and cure AD with adult NSCs.

References

[1] Alzheimer A. Uber einen eigenartigen schweren erkrankungsprozeb der hirnrinde. *Neurologisches cenrealblatt* 23: 1129-1136 (1906).

[2] Caselli RJ, Beach TG, Yaari R, Reiman EM. Alzheimer's disease a century later. *J. Clin. Psychiatry* 67: 1784-1800 (2006).

[3] Jellinger KA. Alzheimer 100 - highlights in the history of Alzheimer research. *J. Neural Transm.* 113: 1603-1623 (2006).

[4] Burns A, Byrne EJ and Maurer K. Alzheimer's disease. *Lancet* 360: 163-165 (2002).

[5] Ferri CP, Prince M, Brayne C, Brodaty H, Fratiglioni L, Ganguli M, et al. Alzheimer's Disease International. Global prevalence of dementia: a Delphi consensus study. *Lancet* 366: 2112-2117 (2005).

[6] Gross CG. Neurogenesis in the adult brain: death of a dogma. *Nat. Rev. Neurosci.* 1: 67-73 (2000).

[7] Taupin P. Neurogenesis in the adult central nervous system. *C. R. Biol.* 329: 465-475 (2006).

[8] Eriksson PS, Perfilieva E, Bjork-Eriksson T, Alborn AM, Nordborg C, Peterson DA et al. Neurogenesis in the adult human hippocampus. *Nat. Med.* 4: 1313-1317 (1998).

[9] Bedard A and Parent A. Evidence of newly generated neurons in the human olfactory bulb. *Brain Res. Dev. Brain Res.* 151: 159-168 (2004).

[10] Curtis MA, Kam M, Nannmark U, Anderson MF, Axell MZ, Wikkelso C, et al. Human neuroblasts migrate to the olfactory bulb via a lateral ventricular extension. *Science* 315: 1243-1249 (2007).

[11] Taupin P and Gage FH. Adult neurogenesis and neural stem cells of the central nervous system in mammals. *J. Neurosci. Res.* 69: 745-749 (2002).

[12] Gage FH. Mammalian neural stem cells. *Science* 287: 1433-1438 (2000).

[13] Parent JM, Tada E, Fike JR, Lowenstein DH. Inhibition of dentate granule cell neurogenesis with brain irradiation does not prevent seizure-induced mossy fiber synaptic reorganization in the rat. *J. Neurosci.* 19: 4508-4519 (1999).

[14] Shors TJ, Miesegaes G, Beylin A, Zhao M, Rydel T, Gould E. Neurogenesis in the adult is involved in the formation of trace memories. *Nature* 410: 372-376 (2001). Erratum in: *Nature* 414, 938 (2001).

[15] Jin K, Peel AL, Mao XO, Xie L, Cottrell BA, Henshall DC, et al. Increased hippocampal neurogenesis in Alzheimer's disease. *Proc. Natl. Acad. Sci. USA* 101: 343-347 (2004).

[16] Duan X, Chang JH, Ge S, Faulkner RL, Kim JY, Kitabatake Y, et al. Disrupted-In-Schizophrenia 1 regulates integration of newly generated neurons in the adult brain. *Cell* 130: 1146-1158 (2007).

[17] Taupin P. Adult neurogenesis and neuroplasticity. *Restor Neurol. Neurosci.* 24: 9-15 (2006).

[18] Taupin P. Adult neurogenesis pharmacology in neurological diseases and disorders. *Exp. Rev. Neurother.* 8: 311-320 (2008).

[19] Lederman RJ. What tests are necessary to diagnose Alzheimer disease? *Cleve Clin. J. Med.* 67: 615-618 (2000).

[20] Dubois B, Feldman HH, Jacova C, DeKosky ST, Barberger-Gateau P, Cummings J, et al. Research criteria for the diagnosis of Alzheimer's disease: revising the NINCDS–ADRDA criteria. *The Lancet Neurology* 6: 734-746 (2007).

[21] Patterson C, Feightner JW, Garcia A, Hsiung GY, MacKnight C, Sadovnick AD. Diagnosis and treatment of dementia: 1. Risk assessment and primary prevention of Alzheimer disease. *CMAJ* 178: 548-556 (2008).

[22] Cankurtaran M, Yavuz BB, Cankurtaran ES, Halil M, Ulger Z, Ariogul S. Risk factors and type of dementia: Vascular or Alzheimer? *Arch Gerontol. Geriatr.* 47: 25-34 (2008).

[23] Terry RD. The pathogenesis of Alzheimer disease: an alternative to the amyloid hypothesis. *J. Neuropathol. Exp. Neurol.* 55: 1023-1025 (1996).

[24] St George-Hyslop PH. Piecing together Alzheimer's. *Sci. Am.* 283: 76-83 (2000).

[25] Brun A, Gustafson L. Distribution of cerebral degeneration in Alzheimer's disease. A clinico-pathological study. *Arch Psychiatr Nervenkr.* 223: 15-33 (1976).

[26] Anderson DH, Talaga KC, Rivest AJ, Barron E, Hageman GS and Johnson LV. Characterization of beta amyloid assemblies in drusen: the deposits associated with aging and age-related macular degeneration. *Exp. Eye Res.* 78: 243-256 (2004).

[27] Kang J, Lemaire HG, Unterbeck A, Salbaum JM, Masters CL, Grzeschik KH, et al. The precursor of Alzheimer's disease amyloid A4 protein resembles a cell-surface receptor. *Nature* 325: 733-736 (1987).

[28] Goldgaber D, Lerman MI, McBride OW, Saffiotti U, Gajdusek DC. Characterization and chromosomal localization of a cDNA encoding brain amyloid of Alzheimer's disease. *Science* 235: 877-880 (1987).

[29] Schellenberg GD, Bird TD, Wijsman EM, Orr HT, Anderson L, Nemens E, et al. Genetic linkage evidence for a familial Alzheimer's disease locus on chromosome 14. *Science* 258: 668-671 (1992).

[30] Serpell LC, Sunde M, Blake CC. The molecular basis of amyloidosis. *Cell Mol. Life Sci.* 53: 871-887 (1997).

[31] Dobson CM. Protein misfolding, evolution and disease. *Trends Biochem. Sci.* 24: 329-332 (1999).

[32] Hardy J, Selkoe DJ. The amyloid hypothesis of Alzheimer's disease: progress and problems on the road to therapeutics. *Science* 297: 353-356 (2002). Erratum in: *Science* 297: 2209 (2002).

[33] Meyer-Luehmann M, Coomaraswamy J, Bolmont T, Kaeser S, Schaefer C, Kilger E, et al. *Science* 313: 1781-1784 (2006).

[34] Billings LM, Oddo S, Green KN, McGaugh JL, LaFerla FM. Intraneuronal Abeta causes the onset of early Alzheimer's disease-related cognitive deficits in transgenic mice. *Neuron* 45: 675-688 (2005).

[35] Fukutani Y, Kobayashi K, Nakamura I, Watanabe K, Isaki K, Cairns NJ. Neurons, intracellular and extra cellular neurofibrillary tangles in subdivisions of the hippocampal cortex in normal ageing and Alzheimer's disease. *Neurosci. Lett.* 200: 57-60 (1995).

[36] Kim H, Jensen CG, Rebhun LI. The binding of MAP-2 and tau on brain microtubules in vitro: implications for microtubule structure. *Ann. N. Y. Acad. Sci.* 466: 218-239 (1986).

[37] Drubin DG, Kirschner MW. Tau protein function in living cells. *J. Cell Biol.* 103: 2739-2746 (1986).

[38] Iqbal K, Grundke-Iqbal I, Smith AJ, George L, Tung YC, Zaidi T. Identification and localization of a tau peptide to paired helical filaments of Alzheimer disease. *Proc. Natl. Acad. Sci. USA* 86: 5646¬5650 (1989).

[39] Wang JZ, Grundke-Iqbal I, Iqbal K. Kinases and phosphatases and tau sites involved in Alzheimer neurofibrillary degeneration. *Eur. J. Neurosci.* 25: 59-68 (2007).

[40] Iqbal K, Alonso AC, Gong CX, Khatoon S, Pei JJ, Wang JZ et al. Mechanisms of neurofibrillary degeneration and the formation of neurofibrillary tangles. *J. Neural. Transm. Suppl.* 53: 169-180 (1998).

[41] Alonso A, Zaidi T, Novak M, Grundke-Iqbal I, Iqbal K. Hyper-phosphorylation induces self-assembly of tau into tangles of paired helical filaments/straight filaments. *Proc. Natl. Acad. Sci. USA* 98: 6923-6928 (2001).

[42] Hardy J and Orr H. The genetics of neurodegenerative diseases. *J. Neurochem.* 97: 1690-1699 (2006).

[43] Hardy J. The genetic causes of neurodegenerative diseases. *J. Alzheimers Dis.* 3: 109-116 (2001).

[44] Schellenberg GD. Genetic dissection of Alzheimer disease, a heterogeneous disorder. *Proc. Natl. Acad. Sci. USA* 92: 8552-8559 (1995).

[45] Thomas P, Fenech M. A review of genome mutation and Alzheimer's disease. *Mutagenesis* 22: 15-33 (2007).

[46] Minghetti L. Role of inflammation in neurodegenerative diseases. *Curr. Opin. Neurol.* 18: 315-321 (2005).

[47] Filipcik P, Cente M, Ferencik M, Hulin I, Novak M. The role of oxidative stress in the pathogenesis of Alzheimer's disease. *Bratisl. Lek. Listy* 107: 384-394 (2006).

[48] Onyango IG, Khan SM. Oxidative stress, mitochondrial dysfunction, and stress signalling in Alzheimer's disease. *Curr. Alzheimer Res.* 3: 339-349 (2006).

[49] Zilka N, Ferencik M, Hulin I. Neuroinflammation in Alzheimer's disease: protector or promoter? *Bratisl. Lek. Listy* 107: 374-383 (2006).

[50] Malinski T. Nitric oxide and nitroxidative stress in Alzheimer's disease. *J. Alzheimers Dis.* 11: 207-218 (2007).

[51] Raber J, Huang Y, Ashford JW. ApoE genotype accounts for the vast majority of AD risk and AD pathology. *Neurobiol. Aging* 25: 641-650 (2004).

[52] Nishimura M, Yu G, St George-Hyslop PH. Biology of presenilins as causative molecules for Alzheimer disease. *Clin. Genet.* 55: 219-225 (1999).

[53] Newman M, Musgrave FI, Lardelli M. Alzheimer disease: amyloi¬dogenesis, the presenilins and animal models. *Biochim. Biophys. Acta* 1772: 285-297 (2007).

[54] Goedert M, Strittmatter WJ, Roses AD. Alzheimer's disease. Risky apolipoprotein in brain. *Nature* 372: 45-46 (1994).

[55] Rogaeva E, Meng Y, Lee JH, Gu Y, Kawarai T, Zou F, et al. The neuronal sortilin-related receptor SORL1 is genetically associated with Alzheimer disease. *Nat. Genet* 39: 168-177 (2007).

[56] Mahley RW. Apolipoprotein E: cholesterol transport protein with expanding role in cell biology. *Science* 240: 622-630 (1988).

[57] Strittmatter WJ, Saunders AM, Schmechel D, Pericak-Vance M, Enghild J, Salvesen GS, et al. Apolipoprotein E: high-avidity binding to beta-amyloid and increased frequency of type 4 allele in late-onset familial Alzheimer disease. *Proc. Natl. Acad. Sci. USA* 90: 1977-1981 (1993).

[58] Corder EH, Saunders AM, Strittmatter WJ, Schmechel DE, Gaskell PC, Small GW, et al. Gene dose of apolipoprotein E type 4 allele and the risk of Alzheimer's disease in late onset families. *Science* 261: 921-923 (1993).

[59] Ma J, Yee A, Brewer HB Jr, Das S, Potter H. Amyloid-associated proteins alpha 1-antichymotrypsin and apolipoprotein E promote assembly of Alzheimer beta-protein into filaments. *Nature* 372: 92¬94 (1994).

[60] Seaman MN. Recycle your receptors with retromer. *Trends Cell Biol.* 15: 68-75 (2005).

[61] Finckh U, von der Kammer H, Velden J, Michel T, Andresen B, Deng A, et al.Genetic association of a cystatin C gene polymorphism with late-onset Alzheimer disease. *Arch Neurol.* 57: 1579¬1583 (2000).

[62] Wakutani Y, Kowa H, Kusumi M, Nakaso K, Yasui K, Isoe-Wada K, et al. A haplotype of the methylenetetrahydrofolate reductase gene is protective against late-onset Alzheimer's disease. *Neurobiol. Aging* 25: 291-294 (2004).

[63] Beyer K, Lao JI, Latorre P, Riutort N, Matute B, Fernández-Figueras MT, et al. Methionine synthase polymorphism is a risk factor for Alzheimer disease. *Neuroreport* 14: 1391-1394 (2003).

[64] Holgado-Madruga M, Emlet DR, Moscatello DK, Godwin AK, Wong AJ. A Grb2-associated docking protein in EGF- and insulin-receptor signalling. *Nature* 379: 560-564 (1996).

[65] Reiman EM, Webster JA, Myers AJ, Hardy J, Dunckley T, Zis¬mann VL, et al. GAB2 Alleles Modify Alzheimer's Risk in APOE varepsilon4 Carriers. *Neuron* 54: 713-720 (2007).

[66] Wragg M, Hutton M, Talbot C. Genetic association between intronic polymorphism in presenilin-1 gene and late-onset Alzheimer's disease. Alzheimer's Disease Collaborative Group. *Lancet* 347: 509-512 (1996).

[67] Migliore L, Testa A, Scarpato R, Pavese N, Petrozzi L, Bonuccelli U. Spontaneous and induced aneuploidy in peripheral blood lymphocytes of patients with Alzheimer's disease. *Hum. Genet* 101: 299-305 (1997).

[68] Migliore L, Botto N, Scarpato R, Petrozzi L, Cipriani G, Bonuccelli U. Preferential occurrence of chromosome 21 malsegregation in peripheral blood lymphocytes of Alzheimer disease patients. *Cytogenet Cell Genet* 87: 41-46 (1999).

[69] Geller LN, Potter H. Chromosome missegregation and trisomy 21 mosaicism in Alzheimer's disease. *Neurobiol. Dis.* 6: 167-179 (1999).

[70] Potter H. Review and hypothesis: Alzheimer disease and Down syndrome-chromosome 21 nondisjunction may underlie both disorders. *Am. J. Hum. Genet* 48: 1192-1200 (1991).

[71] Olson MI, Shaw CM. Presenile dementia and Alzheimer's disease in mongolism. *Brain* 92: 147-156 (1969).

[72] Glenner GG, Wong CW. Alzheimer's disease and Down's syndrome: sharing of a unique cerebrovascular amyloid fibril protein. *Biochem. Biophys. Res. Commun.* 122: 1131-1135 (1984).

[73] Wisniewski HM, Rabe A. Discrepancy between Alzheimer-type neuropathology and dementia in persons with Down's syndrome. *Ann. N. Y. Acad. Sci.* 477: 247-260 (1986).

[74] Roizen NJ, Patterson D. Down's syndrome. *Lancet* 361: 1281-1289 (2003).

[75] Torres EM, Williams BR, Amon A. Aneuploidy: cells losing their balance. *Genetics* 179: 737-746 (2008).

[76] Li J, Xu M, Zhou H, Ma J, Potter H. Alzheimer presenilins in the nuclear membrane, interphase kinetochores, and centrosomes suggest a role in chromosome segregation. *Cell* 90: 917-927 (1997).

[77] Ramírez MJ, Puerto S, Galofré P, Parry EM, Parry JM, Creus A, et al. Multicolour FISH detection of radioactive iodine-induced 17cen-p53 chromosomal breakage in buccal cells from therapeutically exposed patients. *Carcinogenesis* 21: 1581-1586 (2000).

[78] Chen Y, McPhie DL, Hirschberg J, Neve RL. The amyloid precursor protein-binding protein APP-BP1 drives the cell cycle through the S-M checkpoint and causes apoptosis in neurons. *J. Biol. Chem.* 275: 8929-8935 (2000).

[79] Haughey NJ, Nath A, Chan SL, Borchard AC, Rao MS, Mattson MP. Disruption of neurogenesis by amyloid beta-peptide, and perturbed neural progenitor cell homeostasis, in models of Alzheimer's disease. *J. Neurochem.* 6: 1509-1524 (2002).

[80] Klein JA, Ackerman SL. Oxidative stress, cell cycle, and neurodegeneration. *J. Clin. Invest.* 111: 785-793 (2003).

[81] Langley B, Ratan RR. Oxidative stress-induced death in the nervous system: cell cycle dependent or independent? *J. Neurosci. Res.* 77: 621-629 (2004).

[82] Zhu X, Raina AK, Perry G, Smith MA. Alzheimer's disease: the two-hit hypothesis. *Lancet Neurol.* 3: 219-226 (2004).

[83] Jin K, Peel AL, Mao XO, Xie L, Cottrell BA, Henshall DC et al. Increased hippocampal neurogenesis in Alzheimer's disease. *Proc. Natl. Acad. Sci. USA* 101: 343-347 (2004).

[84] Jin K, Galvan V, Xie L, Mao XO, Gorostiza OF, Bredesen DE, et al. Enhanced neurogenesis in Alzheimer's disease transgenic (PDGF-APPSw,Ind) mice. *Proc. Natl. Acad. Sci. USA* 101: 13363¬13367 (2004).

[85] Feng R, Rampon C, Tang YP, Shrom D, Jin J, Kyin M, et al. Deficient neurogenesis in forebrain-specific presenilin-1 knockout mice is associated with reduced clearance of hippocampal memory traces. *Neuron* 32: 911-926 (2001). Erratum in: *Neuron* 33: 313 (2002).

[86] Wen PH, Shao X, Shao Z, Hof PR, Wisniewski T, Kelley K, et al. Overexpression of wild type but not an FAD mutant presenilin-1 promotes neurogenesis in the hippocampus of adult mice. *Neurobiol. Dis.* 10: 8-19 (2002).

[87] Verret L, Jankowsky JL, Xu GM, Borchelt DR, Rampon C. Alzheimer's-type amyloidosis in transgenic mice impairs survival of newborn neurons derived from adult hippocampal neurogenesis. *J. Neurosci.* 27: 6771-6780 (2007).

[88] Zhang C, McNeil E, Dressler L, Siman R. Long-lasting impairment in hippocampal neurogenesis associated with amyloid deposition in a knock-in mouse model of familial Alzheimer's disease. *Exp. Neurol.* 204: 77-87 (2007).

[89] Rodríguez JJ, Jones VC, Tabuchi M, Allan SM, Knight EM, LaFerla FM, et al. Impaired adult neurogenesis in the dentate gyrus of a triple transgenic mouse model of Alzheimer's disease. *PLoS ONE* 3: e2935 (2008).

[90] Donovan MH, Yazdani U, Norris RD, Games D, German DC, Eisch AJ. Decreased adult hippocampal neurogenesis in the PDAPP mouse model of Alzheimer's disease. *J. Comp. Neurol.* 495: 70-83 (2006).

[91] German DC, Eisch AJ. Mouse models of Alzheimer's disease: insight into treatment. *Rev. Neurosci.* 15: 353-369 (2004).

[92] Heo C, Chang KA, Choi HS, Kim HS, Kim S, Liew H, et al. Effects of the monomeric, oligomeric, and fibrillar Abeta42 peptides on the proliferation and differentiation of adult neural stem cells from subventricular zone. *J. Neurochem.* 102: 493-500 (2007).

[93] Taupin P. Adult neurogenesis and neural stem cells in mammals. Nova Science Publishers. New York. First Edition (January 2007).

[94] Klunk WE, Engler H, Nordberg A, Wang Y, Blomqvist G, Holt DP, et al. Imaging brain amyloid in Alzheimer's disease with Pittsburgh Compound-B. *Ann. Neurol* 55: 306-319 (2004).

[95] Small GW, Kepe V, Ercoli LM, Siddarth P, Bookheimer SY, Miller KJ, et al. PET of brain amyloid and tau in mild cognitive impairment. *N. Engl. J. Med.* 355: 2652-2663 (2006).

[96] van Oijen M, Hofman A, Soares HD, Koudstaal PJ, Breteler MM. Plasma Abeta(1-40) and Abeta(1-42) and the risk of dementia: a prospective case-cohort study. *Lancet Neurol.* 5: 655-660 (2006).

[97] Finehout EJ, Franck Z, Choe LH, Relkin N and Lee KH. Cerebrospinal fluid proteomic biomarkers for Alzheimer's disease. *Ann. Neurol.* 61: 120-129 (2007).

[98] Forsberg A, Engler H, Almkvist O, Blomquist G, Hagman G, Wall A, et al. PET imaging of amyloid deposition in patients with mild cognitive impairment. *Neurobiol. Aging* 29: 1456-1465 (2007).

[99] Nordberg A. Amyloid imaging in Alzheimer's disease. *Curr. Opin. Neurol.* 20: 398-402 (2007).

[100] Kawas CH. Clinical practice. Early Alzheimer's disease. *N. Engl. J. Med.* 349: 1056-1063 (2003).

[101] Scarpini E, Scheltens P, Feldman H. Treatment of Alzheimer's disease: current status and new perspectives. *Lancet Neurol.* 2: 539-547 (2003).

[102] Arrieta JL, Artalejo FR. Methodology, results and quality of clinical trials of tacrine in the treatment of Alzheimer's disease: a systematic review of the literature. *Age Ageing* 27: 161-179 (1998).

[103] Wilkinson DG, Francis PT, Schwam E, Payne-Parrish J. Cholinesterase inhibitors used in the treatment of Alzheimer's disease: the relationship between pharmacological effects and clinical efficacy. *Drugs Aging* 21: 453-478 (2004).

[104] [104] Aisen PS. The development of anti-amyloid therapy for Alzheimer's disease: from secretase modulators to polymerisation inhibitors. *CNS Drugs* 19: 989-996 (2005).

[105] Creeley C, Wozniak DF, Labruyere J, Taylor GT, Olney JW. Low doses of memantine disrupt memory in adult rats. *J. Neurosci.* 26: 3923-3932 (2006).

[106] McShane R, Areosa Sastre A, Minakaran N. Memantine for dementia. *Cochrane Database Syst. Rev.* 2: CD003154 (2006).

[107] Estrada LD, Soto C. Disrupting beta-amyloid aggregation for Alzheimer disease treatment. *Curr. Top Med. Chem.* 7: 115-126 (2007).

[108] Janus C. Vaccines for Alzheimer's disease: how close are we? *CNS Drugs* 17: 457-474 (2003).

[109] Solomon B. Intravenous immunoglobulin and Alzheimer's disease immunotherapy. *Curr. Opin. Mol. Ther.* 9: 79-85 (2007).

[110] Rehen SK, McConnell MJ, Kaushal D, Kingsbury MA, Yang AH, Chun J. Chromosomal variation in neurons of the developing and adult mammalian nervous system. *Proc. Natl. Acad. Sci. USA* 98: 13361-13366 (2001).

[111] Rehen SK, Yung YC, McCreight MP, Kaushal D, Yang AH, Almeida BS, et al. Constitutional aneuploidy in the normal human brain. *J. Neurosci.* 25: 2176-2180 (2005).

[112] Alvarez-Dolado M, Pardal R, Garcia-Verdugo JM, Fike JR, Lee HO, Pfeffer K, et al. Fusion of bone-marrow-derived cells with Purkinje neurons, cardiomyocytes and hepatocytes. *Nature* 425: 968-973 (2003).

[113] Yang Y, Geldmacher DS, Herrup K. DNA replication precedes neuronal cell death in Alzheimer's disease. *J. Neurosci.* 21: 2661-2668 (2001).

[114] Busser J, Geldmacher DS, Herrup K. Ectopic cell cycle proteins predict the sites of neuronal cell death in Alzheimer's disease brain. *J. Neurosci.* 18; 2801-2807 (1998).

[115] Vincent I, Rosado M, Davies P. Mitotic mechanisms in Alzheimer's disease? *J. Cell Biol.* 132: 413-425 (1996).

[116] Herrup K, Neve R, Ackerman SL, Copani A. Divide and die: cell cycle events as triggers of nerve cell death. *J. Neurosci.* 24: 9232-9239 (2004).

[117] Yang Y, Mufson EJ, Herrup K. Neuronal cell death is preceded by cell cycle events at all stages of Alzheimer's disease. *J. Neurosci.* 23: 2557-2563 (2003).

[118] Herrup K, Arendt T. Re-expression of cell cycle proteins induces neuronal cell death during Alzheimer's disease. *J. Alzheimers Dis.* 4: 243-247 (2002).

[119] Morshead CM, van der Kooy D. Postmitotic death is the fate of constitutively proliferating cells in the subependymal layer of the adult mouse brain. *J. Neurosci.* 12: 249-256 (1992).

[120] Cameron HA, McKay RD. Adult neurogenesis produces a large pool of new granule cells in the dentate gyrus. *J. Comp. Neurol.* 435: 406-417 (2001).

[121] Kempermann G, Kuhn HG, Gage FH. More hippocampal neurons in adult mice living in an enriched environment. *Nature* 386: 493¬495 (1997).

[122] Miller MW and Nowakowski RS. Use of bromodeoxyuridine-immunohistochemistry to examine the proliferation, migration and time of origin of cells in the central nervous system. *Brain Res.* 457: 44-52 (1988).

[123] del Rio JA and Soriano E. Immunocytochemical detection of 5'-bromodeoxyuridine incorporation in the central nervous system of the mouse. *Brain Res. Dev. Brain Res.* 49: 311-317 (1989).

[124] Nowakowski RS and Hayes NL. New neurons: extraordinary evidence or extraordinary conclusion? *Science* 288: 771 (2000).

[125] Gould E and Gross CG. Neurogenesis in adult mammals: some progress and problems. *J. Neurosci.* 22: 619-623 (2002).

[126] Rakic P. Adult neurogenesis in mammals: an identity crisis. *J. Neurosci.* 22: 614-618 (2002).

[127] Taupin P. BrdU Immunohistochemistry for Studying Adult Neurogenesis: paradigms, pitfalls, limitations, and validation. *Brain Res. Rev.* 53: 198-214 (2007).

[128] Taupin P. Protocols for Studying Adult Neurogenesis: Insights and Recent Developments. *Reg. Med.* 2: 51-62 (2007).

[129] Poduslo JF, Curran GL, Wengenack TM, Malester B, Duff K. Permeability of proteins at the blood-brain barrier in the normal adult mouse and double transgenic mouse model of Alzheimer's disease. *Neurobiol. Dis.* 8: 555-567 (2001).

[130] Desai BS, Monahan AJ, Carvey PM, Hendey B. Blood-brain barrier pathology in Alzheimer's and Parkinson's disease: implications for drug therapy. *Cell Transplant* 16: 285-299 (2007).

[131] Taupin P. The therapeutic potential of adult neural stem cells. *Curr. Opin. Mol. Ther.* 8: 225-231 (2006).

[132] Mitsiadis TA, Barrandon O, Rochat A, Barrandon Y, De Bari C. Stem cell niches in mammals. *Exp. Cell Res.* 313: 3377-3385 (2007).

[133] Ninkovic J, Götz M. Signalling in adult neurogenesis: from stem cell niche to neuronal networks. *Curr. Opin. Neurobiol.* 17: 338-344 (2007).

[134] Taupin P. Adult neural stem cells, neurogenic niches and cellular therapy. *Stem Cell Reviews* 2: 213-220 (2006).

[135] Chin VI, Taupin P, Sanga S, Scheel J, Gage FH, Bhatia SN. Micro-fabricated platform for studying stem cell fates. *Biotechnol. Bioeng.* 88: 399-415 (2004).

[136] Brown AB, Yang W, Schmidt NO, Carroll R, Leishear KK, Rainov NG, et al. Intravascular delivery of neural stem cell lines to target intracranial and extracranial tumors of neural and non-neural ori¬gin. *Hum. Gene Ther.* 14: 1777-1785 (2003).

[137] Pluchino S, Quattrini A, Brambilla E, Gritti A, Salani G, Dina G, et al. Injection of adult neurospheres induces recovery in a chronic model of multiple sclerosis. *Nature* 422: 688-694 (2003).

[138] Taupin P. Consideration of adult neurogenesis from basic science to therapy. *Med. Sci. Monit.* 11: LE16-LE17 (2005).

Chapter VI

Adult Neurogenesis Pharmacology in Neurological Diseases and Disorders

Abstract

With the confirmation that neurogenesis occurs in the adult brain and neural stem cells reside in the adult CNS, the focus of research has now shifted to the understanding of the function of newborn neuronal cells in the adult brain, and particularly in the pathologies of the nervous system. Neurogenesis has been reported to be modulated in a broad range of pathological conditions, including neurological diseases and disorders. More strikingly, studies have revealed that drugs currently used to treat neurological diseases and disorders, such as Alzheimer's disease and depression, increase adult neurogenesis, which may mediate their activities. However, some of these studies are the source of debates and controversies, and remain to be confirmed. Hence, the role and contribution of newly generated neuronal cells in neurological diseases and disorders, as well as the effect of drugs on adult neurogenesis and its significance, remain to be elucidated and understood. This shows that adult neurogenesis is not only important for our understanding of development and therapy, but also for the physiopathology of the CNS and its pharmacology.

Key Issues

- In mammals, neurogenesis occurs in the adult brain and neural stem cells reside in the adult CNS.
- The function of newborn neuronal cells in the physiology and pathology of the adult brain remains to be elucidated.
- Adult neurogenesis is modulated in neurological diseases and disorders, and by drugs used to treat neurological diseases and disorders, such as Alzheimer's disease and depression.

- The role, significance and mechanisms of the modulation of adult neurogenesis in the etiology of neurological diseases and disorders, and the activities of drugs used to treat neurological diseases and disorders, remain to be elucidated.
- The modulation of adult neurogenesis in neurological diseases and disorders, and by drugs used to treat neurological diseases and disorders, remain to be further evaluated and confirmed.
- The modulation of adult neurogenesis by drugs, used to treat neurological diseases and disorders, may lead to new drug design and strategies to treat neurological diseases and disorders.
- Anti-inflammatory treatments offer new perspectives for the treatment of neurological diseases and disorders.

Introduction

Most nerve cells in the adult mammalian CNS are postmitotic and differentiated cells. They are born from primordial stem cells during development. It was believed that the adult brain was devoid of stem cells, hence lacks the capacity to generate new nerve cells and regenerate after injury [1]. Seminal studies in the1960s [2, 3] and studies, mostly carried out in the 1980s and 1990s [4], have revealed and confirmed that contrary to a long-held dogma, neurogenesis occurs in the adult brain and neural stem cells (NSCs) reside in the adult CNS of mammals [5]. The confirmation that neurogenesis occurs in the adult mammalian brain has tremendous consequences for development and therapy, but also for our understanding of the physiology and pathology of the CNS.

This review begins by introducing basic NSC biology. Then, the diseases in which NSCs are affected are discussed. Subsequently, we discuss the drugs used to treat these disorders. Finally, we discuss how these drugs alter NSCs.

Neurogenesis in the Adult Brain

Neurogenesis occurs throughout adulthood in mammals [4]. It occurs primarily in two regions of the adult brain, the anterior portion of the subventricular zone (SVZ) and the dentate gyrus (DG) of the hippocampus, including in humans [6, 7]. In the SVZ, newborn progenitor cells migrate to the olfactory bulb, trough the rostro–migratory stream, where they differentiate into interneurons [7–9]. In the DG, newborn neuronal cells in the subgranular zone (SGZ) migrate to the granule cell layer, where they differentiate into granule-like cells [10] and extend axonal projections to region CA3 of Ammon's horn [11, 12]. Newborn neuronal cells in the DG and SVZ establish functional connections with neighboring cells [13–16]. Adult neurogenesis may reproduce processes similar to those that occur during development, to integrate newborn neuronal cells in the hippocampal network [17].

The number of newborn neuronal cells generated in the adult brain is relatively low, particularly in the DG. A total of 0.1 and 0.004% of the granule cell population are generated per day in the DG of young adult rodents and adult macaque monkeys, respectively [18–20].

Some progenitor cells survive for extended period of time, at least 2 years in the DG of humans [6]. Adult neurogenesis is affected by various conditions, such as the environment, trophic factors, hormones and drug treatments, various physio- and pathological processes, such as aging, diseases, disorders and injuries, as well as genetic background [21].

It is postulated that newborn neuronal cells originate from stem cells in the adult brain. In support to this contention, self-renewing multipotent neural progenitor and stem cells have been isolated and characterized in vitro from various regions ofbthe adult CNS [4, 22, 23]. In the adult brain, populations of astrocytes and ependymocytes have been identified and proposed as candidates for stem cells in the SVZ and DG [24–26]. NSCs are the self-renewing multipotent cells that generate the main phenotypes of the nervous system [27]. Despite being characterized in vitro and in situ, NSCs are still elusive cells in the adult CNS and remain to be unequivocally identified and characterized in vitro and in vivo [28–32].

Hence, functional neurogenesis occurs in the adult brain. Newborn neuronal cells in the adult brain may replace nerve cells born during development. Although the involvement and contribution of newborn neuronal cells in the adult brain remain to be elucidated, the confirmation that neurogenesis occurs in the adult brain suggests that it has the potential for self-repair. The existence of NSCs in the adult CNS has tremendous consequences for our understanding of brain functioning and for cellular therapy. Cellular therapy may involve the stimulation of endogenous neural progenitor or stem cells, or the transplantation of neural progenitor and stem cells to repair the degenerated or injured nervous system [33].

Neurogenesis in Neurological Diseases and Disorders

Alzheimer's Disease

Neurogenesis has been reported to be enhanced in the hippocampus of patients with Alzheimer's disease (AD) [34]. Studies from autopsies reveal that the expression of markers of immature neuronal cells, such as doublecortin and polysialylated nerve cell adhesion molecule, is increased in hippocampal regions, particularly the DG, of the brains of AD patients. In animal models of AD, neurogenesis is enhanced in the DG of transgenic mice that express the Swedish and Indiana amyloid precursor protein (APP) mutations, a mutant form of human APP [35]. It is decreased in the DG and SVZ of knockout mice or mice deficient for presenilin-1 (PS-1) and APP [36, 37]. It is decreased in the DG of PDAPP transgenic mice, a mouse model of AD with age-dependent accumulation of amyloid protein [38]. It is also decreased in transgenic mice overexpressing familial AD variants of APP or PS-1 [39] and in mice with targeted mutations in APP and/or PS-1 [40].

The late-onset form of AD (LOAD) is the most common form of the disease. It is not inherited and generally develops in people over 65 years of age [41]. It is also referred to as sporadic AD. LOAD is the most probable form of the disease among the 14 patients with a clinical diagnosis of AD that were included in the study by Jin and collaborators [34]. Early-onset AD is a very rare form of the disease, referred to as familial AD. It is primarily genetic of origin and strikes younger people, under the age of 65 years [42]. Gene mutations such as APP and PS-1, have been identified as causes of the genetic form of AD [42].

The discrepancies between studies in humans and animal models, with regard to the modulation of adult neurogenesis, could be explained by the limitation of the transgenic animal models as representatives of complex diseases to study adult phenotypes, such as adult neurogenesis [43, 44]. In particular, mutant or deficient mice for single genes, such as PS-1 and APP, may not fully reproduce the features of AD. In addition, in animal models with accumulation of amyloid protein, such as in PDAPP transgenic mice, the decrease of neurogenesis could originate from the deposit of amyloid or its toxic intermediates that may inhibit neurogenesis [38]. Overall, however, although these results show that neurogenesis is enhanced in the brains of patients with AD, the effects of the pathology of AD on neurogenesis remain to be elucidated.

Depression

Stress (an important causal factor in precipitating episodes of depression) and glucocorticoids (stress-related hormones) decrease neurogenesis in the hippocampus. It does not decrease neurogenesis in the SVZ in animal studies, including in non-human primates [45–47]. A post-mortem study reveals that adult neurogenesis is not altered in the hippocampus of patients with major depression [48]. However, this study included only six subjects, many of whom were on antidepressant medication. Hence, further studies need to be performed to confirm whether neurogenesis is altered in human patients with depression. These results show that the relationship between adult neurogenesis and depression remains to be further investigated.

Epilepsy

Neurogenesis is enhanced in the DG and SVZ of animal models of epilepsy, such as after pilocarpine treatment [49]. After pilocarpine treatment, ectopic granule-like cells in the hilus are labeled for bromodeoxyuridine (BrdU). BrdU is a thymidine analog that incorporates DNA of dividing cells during the S-phase of the cell cycle and is used for birthdating and monitoring cell proliferation [50]. However, BrdU is not a marker of cell proliferation and neurogenesis, it is marker of DNA synthesis [51]. Mossy fiber (MF)-like processes immunostained for TOAD-64, a marker for newly generated neuronal cells, are also detected in the granule cell layer of the stratum oriens of CA3 and the inner molecular layer of the DG in rodents [49]. This shows that ectopic granule-like cells originate from newborn neuronal cells and MF remodeling derives from newborn granule cells rather than from pre-existing mature dentate granule cells. Because newborn neuronal cells are mitogenic, they are sensitive to radiation [52]. Low-dose, whole-brain, X-ray irradiation in adult rats after pilocarpine treatment decreases neurogenesis but does not prevent the induction of recurrent seizures, nor prevent seizure-induced ectopic granule-like cells and MF sprouting [53]. These data show that neurogenesis is enhanced in the DG and SVZ and that seizure-induced ectopic granule-like cells and MF sprouting arise not only from newborn neuronal cells, but also from mature dentate granule cells in animal models of epilepsy. These data also suggest that neurogenesis is not critical to epileptogenesis.

In humans, adult neurogenesis has been characterized in patients suffering from chronic temporal lobe epilepsy (TLE). Immunohistochemistry and confocal microscopy analysis of biopsies for markers of the cell cycle, neural progenitor cells and neuronal differentiation, such as Ki-67, nestin and α-tubulin, respectively, reveal an increase in nestin-immunoreactive cells within the hilus and DG. It also reveals an increased Ki-67 proliferation index of nestin-immunoreactive cells in the molecular layer of the DG in the brains of patients with TLE [54]. This suggests that cell proliferation and neurogenesis are increased in the DG of patients with epilepsy. Another study, using immunocytochemistry, quantitative western blot and real-time reverse-transcriptase PCR in surgically resected hippocampi from TLE patients, reveals that the expression of the cell proliferation marker MCM2, and the neuronal differentiation marker doublecortin, decreased significantly with age in controls and in TLE patients; this occurred independently of the degree of granule cell dispersion in the DG. These results indicate that epileptic activity does not stimulate neurogenesis in the human DG, and that granule cells dispersion probably does not result from newly generated granule cells, but rather from an abnormal migration of mature granule cells [55], as observed in animal models of epilepsy [53]. In summary, neurogenesis is enhanced in the DG and SVZ in animal models of epilepsy and in human patients with epilepsy. Seizure-induced ectopic granule-like cells and MF sprouting arise primarily from mature dentate granule cells. The modulation of neurogenesis in human patients with epilepsy remains to be further evaluated and confirmed.

Huntington's Disease

Immunohistochemistry and confocal microscopy analysis of autopsies for markers of the cell cycle and neuronal differentiation, such as proliferating cell nuclear antigen and α-tubulin, reveal that cell proliferation and neurogenesis are increased in the SVZ of patients with Huntington's disease (HD) [56]. In adult R6/1 transgenic mouse model of HD, neurogenesis is decreased in the DG [57]. After quinolinic acid striatal lesioning of adult brain, an experimental model of HD, neurogenesis is enhanced in the SVZ, leading to the migration of neuroblasts and formation of new neuronal cells in damaged areas of the striatum [58], as observed in the brains of HD patients [56].

These data provide evidence that adult neurogenesis is enhanced in the SVZ of patients with HD. It also shows that neural progenitor cells from the SVZ migrate toward the site of degeneration in HD. Data from R6/1 transgenic mice are difficult to interpret in the context of adult neurogenesis in HD, as mutated forms of Huntingtin affect brain development [59]. This could underlie the decrease of adult neurogenesis reported in R6/1 transgenic mice.

Parkinson's Disease

One study reported that the rate of neurogenesis, measured by BrdU labeling, is stimulated in the substantia nigra (SN), following lesioning induced by a systemic dose of 1-methyl-4-phenyl-1,2,3,6-tetrahydropyridine (MPTP) [60]. Other studies reported no evidence of new dopaminergic neurons in the SN of 6-hydroxydopamine (6-OHDA)-lesioned hemi-parkinsonian rodents [61, 62]. Neurogenesis in the adult SN has been the source of debates and contróversies, and remains to be confirmed. There are also controversies over the generation of new nerve cells in the striatum [60–62].

Besides these controversies, there is evidence that dopaminergic neuronal cells of the striatum exert a neurogenic activity on neural progenitor cells of the SVZ and SGZ [63]. Experimental depletion of dopamine in rodents decreases precursor cell proliferation in the SVZ and SGZ [64]. This decrease in proliferation is restored after administration of selective agonists of D2-like receptors. In adult mice, destruction of the dopaminergic neurons in the SN and ventral tegmental area in a 6-OHDA model of Parkinson's disease (PD) reduced the number of proliferating neural precursors in the SVZ of the anterior lateral ventricle by approximately 40% [65]. The generation of neural progenitor cells in the SVZ and SGZ is impaired in the brains of individuals with PD [64]. Hence, dopaminergic innervation and dopamine play a role in the regulation of endogenous neurogenesis in the adult mammalian brain [66].

Schizophrenia

Autopsy studies revealed that neurogenesis is decreased in the DG of patients with schizophrenia [48]. The number of BrdU-positive cells decreased by 23% in the SGZ of the DG 24 h after repeated injections of phencyclidine in rats, an experimental model of schizophrenia [67]. The level of newly generated cells returns to control level 1week after injection in the SGZ. These results show that neurogenesis is decreased in the DG of patients with schizophrenia.

Neurogenesis and the Etiology of Neurological Diseases and Disorders

Neurogenesis is modulated in a broad range of neurological diseases and disorders (table 1). Although some of these results remain to be further evaluated and confirmed, this suggests that adult neurogenesis may be involved in the etiology of these diseases. However, the involvement of adult neurogenesis in the etiology of neurological diseases and disorders remain to be established.

Adult neurogenesis is increased after experimental brain injuries, induced by neuronal damage or insult. An increase in neurogenesis after experimental injuries was first reported in1997 by Gould and Tanapat after excitotoxic and mechanical lesions in the dentate granule cell layer of adult rats [68]. The authors observed an increase in proliferating cells in the SGZ on the side of the lesion compared with the unlesioned side 24 h after surgery. An induction of neurogenesis at the site of degeneration, in the thalamus, was also reported after targeted apoptosis of cortical pyramidal neurons by photolytic lesions in mice [69]. These studies were the first to report that following neuronal damage or insults new neuronal cells are generated at the sites of injury.

Table 1. Regulation of adult neurogenesis in neurological diseases and disorders

Disease/Model	Regulation	Reference #
Alzheimer's disease		
Autopsies	increase	[34]
Transgenic mice Swedish and Indiana APP mutations	increase	[35]
Knock-out/deficient mice for presenilin-1 (PS-1) and APP	decrease	[36, 37]
Transgenic mice PDAPP	decrease	[38]
Mice overexpressing Alzheimer's disease variants of APP or PS-1	decrease	[39, 40]
Depression		
Stress	decrease	[45-47]
Autopsies	not altered	[48]
Epilepsy		
Animal models (e.g., pilocarpine treatment)	increase	[49, 53]
Biopsies	increase	[54]
Huntington's disease		
Autopsies	increase	[56]
R6/1 transgenic mouse model of Huntington's disease	decrease	[57]
Quinolinic acid striatal lesion	increase	[58]
Parkinson's disease		
MPTP lesion	increase	[60]
6-hydroxydopamine lesion	not altered	[61, 62]
Schizophrenia		
Autopsies	decrease	[48]
Phencyclidine injections	decrease	[67]

Adult neurogenesis is modulated in a broad range of neurological diseases and disorders. The role, significance and mechanisms of the modulation of adult neurogenesis in the etiology of neurological diseases and disorders, and the activities of drugs used to treat neurological diseases and disorders, remain to be elucidated.
APP: Amyloid precursor protein; MPTP: 1-methyl-4-phenyl-1,2,3,6-tetrahydropyridine; PS-1: Presenilin-1.

This suggests that the increased neurogenesis in many of these illnesses could result from damage or stimulation induction of neurogenesis. This indicates that neurogenesis may be a result, rather than a cause, of the illness. In support of this contention, data presented provide a strong argument against a critical role of adult neurogenesis in the etiology of neurological diseases and disorders, particularly in epileptogenesis. Hence, increased hippocampal neurogenesis may not be a primary effector to epileptogenesis or another disease. However, a lot of research remains to be done before we can draw a firm conclusion on the contribution and involvement of adult neurogenesis in the etiology and pathology of neurological diseases and disorders.

Pharmacology of Adult Neurogenesis

The activity of drugs used to treat AD and depression on adult neurogenesis has been characterized (table 2). Galantamine, an acetylcholinesterase (AChE) inhibitor and memantine, an NMDA-glutamate receptor antagonist, both used to treat AD, increase neurogenesis in the DG and SVZ of adult rodents by 26–45%, as revealed by BrdU labeling [70]. Chronic administration of antidepressants, such as the selective serotonin reuptake inhibitor (SSRI) fluoxetine, increases neurogenesis in the DG but not the SVZ of adult rats and non-human primates [71–73]. Agomelatine, a melatonergic agonist and serotoninergic antagonist, defining a new class of antidepressant, increases adult hippocampal neurogenesis in rodents [74]. These data suggest that adult neurogenesis may contribute to the activities of drugs used to treat AD and depression. Santarelli et al. (2003) further reported that X-irradiation of the hippocampal region, but not other brain regions, such as the SVZ or the cerebellar region, prevents the behavioral effect of SSRIs, such as fluoxetine, in adult mice (129SvEvTac) [75]. In these mice the activity of SSRIs was reported to be mediated by the serotonin (5-hydroxytryptamine [5-HT])1A receptor. Hence, it is proposed that adult neurogenesis mediates the activities of antidepressants, particularly of SSRIs, and that the activity of SSRIs on adult neurogenesis is mediated by 5-HT receptors. These results, together with the observation that stress, an important causal factor in precipitating episodes of depression, decreases hippocampal neurogenesis [46], led to anew theory of depression. It is proposed that the waning and waxing of hippocampal neurogenesis are important causal factors in the precipitation and recovery from episodes of clinical depression [76].

The involvement of adult neurogenesis in the activity of antidepressants, particularly SSRIs, has been further defined by other studies. In BALB/cJ mice, SSRIs, such as fluoxetine, produce their activities independently of neurogenesis. In these mice, the activities of SSRIs were reported not to be mediated by the 5-HT1A receptor [77]. It is proposed that SSRIs produce antidepressant-like effects via distinct mechanisms in different mouse strains. The mechanism of activity of SSRIs, mediating their antidepressant-like activity, in BALB/cJ mice remains to be characterized. N-[3-(1-{[4-(3,4-difluorophenoxy)-phenyl]methyl}(4-piperidyl))-4-methylphenyl]-2-methylpropanamide (SNAP 94847) , an antagonist of the melanin-concentrating hormone receptor (MCHR1), elicit anxiolytic/antidepressant activities [78]. SNAP 94847 stimulates the proliferation of progenitor cells in the DG, but its activity is unaltered in mice in which neurogenesis was suppressed by X-irradiation [78]. Altogether, these data suggest that antidepressants, including SSRIs, produce their activities via distinct mechanisms, some independent of adult neurogenesis. Hence, the role and significance of the increased neurogenesis in the activity of drugs used to treat neurological diseases and disorders, particularly depression, remain to be fully understood and further evaluated.

Table 2. Samples of studies on adult neurogenesis and their paradigms

Family	Drug	Dose administered	Ref.
Alzheimer's drugs			
AChE inhibitor	Tacrine	5 mg/kg for 14 days	[70]
AChE inhibitor	Galantamine	5 mg/kg for 14 days	[70]
NMDA receptor Antagonist	Memantine	7.5 mg/kg for 14 days	[70]
Antidepressants			
SSRI	Fluoxetine	5 mg/kg for 1, 5, 14 or 28 days	[71]
MAOI	Tranylcypromine	7.5 mg/kg 7 days, 10 mg/kg 14 days	[71]
SNRI	Reboxetine	20 mg/kg 2x per day for 21 days	[71]
SSRI	Fluoxetine	10 mg/kg/day for 5, 11 or 28 days	[75]
TCA	Imipramine	20 mg/kg/day for 5 or 28 days	[75]
TCA	Desipramine	20 mg/kg/day for 5 or 28 days	[75]
SSRI	Fluoxetine	18 mg/kg/day for 28 days	[78]
TCA	Desipramine	20 mg/kg/day for 28 days	[78]

Experiments were performed in rodents. Drugs were administered intraperitoneally [70], except for memantine, intragastric. Drugs were administered intraperitoneally [71]. Drugs were administered orally [75, 78].

AChE: Acetylcholinesterase; MAOI: Monoamine oxidase inhibitor; SNRI: Selective norepinephrine reuptake inhibitor; SSRI: Selective serotonin reuptake inhibitor; TCA: Tricyclic antidepressant.

In PD, dopaminergic innervation and dopamine plays a role in the regulation of endogenous neurogenesis in the adult mammalian brain [66]. Neural progenitor cells of the adult SVZ express dopamine receptors and receive dopaminergic afferents.

Dopaminergic projections from the midbrain to the neostriatum and nucleus accumbens overlaps with the most active region of neurogenesis in the adult brain, the SVZ of the anterior lateral ventricle [79]. This would underlie the activity of dopaminergic neurons and dopamine on endogenous neurogenesis [65]. This identifies dopamine as one of the few known endogenous regulators of adult neurogenesis, with implications for the potential use of endogenous neural precursors in cell replacement strategies for PD. These results raise the possibilities that dopamine receptors and, therefore, agonists may be useful to harness endogenous neurogenesis in cell replacement strategies for PD.

Discussion

Five prescription drugs are currently approved to treat patients diagnosed with AD: tacrine, donepezil, galantamine andrivastigmine, AChE inhibitors; and memantine (a NMDA-glutamate receptor antagonist) [80–82]. These drugs produce improvements in cognitive and behavioral symptoms of AD. AChE inhibitors are thought to improve cognitive functions by enhancing cholinergic neurotransmission in affected brain regions of AD. SSRIs, such as fluoxetine, monoamine oxidase inhibitors, such as tranylcypromine, selective norepinephrine reuptake inhibitors (SNRIs), such as reboxetine and phosphodiesterase-IV inhibitors are currently prescribed for the treatment of depression [83, 84]. SSRIs may produce their therapeutic effects by increasing brain levels of 5-HT, a neurotransmitter implicated in the modulation of mood and anxiety-related disorders [85, 86]. Among the 5-HT receptor subtypes, the 5-HT1A receptor has been prominently implicated in the modulation of mood and anxiety-related disorders [87]. The notion that the activity of these drugs may act on or even be mediated through adult neurogenesis is provocative. However, there are currently a lot of debates and controversies surrounding adult neurogenesis and NSC research. Hence, the contribution of adult neurogenesis and NSCs to the etiology of neurological diseases and disorders, as well as the activity, function and significance of drugs used to treat neurological diseases and disorders, particularly AD and depression, remain to be further established and elucidated.

Limitations and Pitfalls of the Use of BrdU Labeling for Studying Neurogenesis

Most of the studies conducted in animal models of neurological diseases and disorders use BrdU labeling as a paradigm to study adult neurogenesis. Thymidine analogs, such as BrdU, are incorporated by the DNA of dividing cells during the S-phase of the cell cycle. As such, they are used for birthdating and monitoring cell proliferation [50]. However, their use and particularly the use of BrdU for studying neurogenesis, is not without limitations and pitfalls [51, 88]. BrdU is a toxic and mutagenic substance: it triggers cell death, the formation of teratomes, alters DNA stability, lengthens the cell cycle, and has mitogenic, transcriptional and translational effects on cells that incorporate it, all of which have profound consequences on neurogenesis [89–91]. As a thymidine analog, BrdU is not a marker for cell proliferation, but a marker for DNA synthesis. Therefore, studying neurogenesis with BrdU requires distinguishing cell proliferation and neurogenesis from other events involving DNA synthesis, such as DNA repair, abortive cell cycle re-entry and gene duplication, without cell division [51, 92]. A variety of treatments and conditions affect the permeability of the BBB and/or alter the flow of blood to the brain. The BBB and/or cerebral flow are particularly affected after glucocorticoid treatments [93], kainic acid treatments [94], stress [95] and in various neurological diseases and disorders, such as AD and epilepsy [96, 97]. In these conditions, an increase in BrdU labeling in the brain could originate from an increase in BrdU uptake rather than an increase in cell proliferation and neurogenesis [51, 92]. Hence, data involving the use of BrdU, as paradigm for studying adult neurogenesis in neurological diseases and disorders, and after drug treatments, must be carefully assessed and analyzed.

Alzheimer's Disease

The adult brain contains a substantial fraction of aneuploid cells (5–7% in mice) [98, 99]. These cells may originate from abortive cell cycle re-entry or DNA duplication, without cell division and cell fusion [51]. Proteins of the cell cycle, such as cyclin B, a marker of the phase G2, are expressed in neurons in regions in which degeneration occurs, such as the hippocampus, in brains of patients with AD [100]. Some at-risk neurons in the brains of AD patients are also aneuploids [101]. A total of 4–10% of neurons in regions of degeneration were reported to express cell cycle proteins, such as cyclin B or tetraploids [100, 101]. The origin and fate of these cells is yet to be determined, but these data suggest that abortive cell cycle re-entry and DNA duplication, without cell proliferation, occurs at a relatively high frequency in regions of neurodegeneration, such as the hippocampus. Hence, some of the data observed by means of BrdU labeling may not represent adult neurogenesis, but rather labeled nerve cells that may have entered the cell cycle and undergone DNA replication, but did not complete the cell cycle [102]. Recently, a study has reported that an increase in proliferation of glial and vascular cells, but not neurogenesis, occurs in the hippocampus in animal models of AD [103]. Adult neurogenesis in AD must therefore be re-examined in light of these data.

Depression

Clinical MRI and post-mortem studies in depressive patients, as well as in animal studies, reveal that chronic stress and depression result in atrophy of the hippocampus [104–107], and that these effects can be reversed by antidepressants [108]. This suggests that the atrophy observed in the hippocampus of patients with depression could be associated with a decrease in adult neurogenesis. In support of this contention, stress and glucocorticoids decrease hippocampal neurogenesis [45–47]. However, a link between adult neurogenesis and depression remains to be established, as post-mortem studies do not reveal any increase in neurogenesis in the hippocampus of patients with depressive disorders [48].

There are also controversies and debates over the role of the hippocampus and adult neurogenesis in the activity of antidepressants [109–111]. On the one hand, inhibition of 5-HT synthesis and selective lesions of 5-HT neurons are associated with a decrease in the number of newborn cells in the DG and SVZ [112]. On the other hand, there are limitations over validity of animal models of depression and X-irradiation, as representative of the human disorder and as a tool to study adult neurogenesis, respectively. Some studies do not report loss of nerve cells, atrophy or decrease of hippocampal volume in patients with depression or in animal models of depression [113–115]. The hippocampus may not be the brain region primarily involved in depressive episodes, as other areas may play a critical role in depression [116]. In conclusion, the involvement of adult neurogenesis in the activity of antidepressants remains to be established.

Because of these limitations, pitfalls and controversies, establishing the mechanisms underlying drug activities would contribute to the understanding of the relationship between adult neurogenesis, neurological diseases and disorders, and drug activities. Particularly, recent studies associate inflammatory responses to the etiology of neurological diseases and disorders, particularly AD [117, 118]. Neuroinflammation inhibits neurogenesis in the adult hippocampus [119, 120]. Although the mechanisms of such regulation has yet to be determined on the cellular level, neurological diseases and disorders are associated with microglia activation [121], a component of the inflammation reaction known to impair hippocampal neurogenesis in adult rats [119, 120]. On the molecular level, substances released by the immune cells, such as interleukin [122] and nitric oxide [123], negatively regulate adult neurogenesis. Hence, neuroinflammation may contribute to the effects of neurological diseases and disorders on adult neurogenesis. This further supports strategies for the treatments of neurological diseases and disorders with anti-inflammatory drugs.

Expert Commentary and Five-Year View

Adult neurogenesis is modulated in neurological diseases and disorders, and may contribute to the activities of drugs used for the treatment of neurological diseases and disorders, particularly AD and depression. The role, significance and mechanisms of the modulation of adult neurogenesis in neurological diseases and disorders, and in mediating drug activities remain to be elucidated and established. Despite evidence of the involvement of adult neurogenesis in the etiology of neurological diseases and disorders and drug activities, these data remain to be further evaluated and confirmed. In particular, limitations and pitfalls over the use of BrdU for studying neurogenesis may have led to some misinterpretation of data.

The modulation of neurogenesis in neurological diseases and disorders may contribute to regenerative processes, as well neuronal plasticity. The contribution of adult neurogenesis to drugs may hold the key for the understanding of their activities, as well as to design new drugs and strategies to treat neurological diseases and disorders. To this aim, neuroinflammation may open new opportunities for the treatment of these diseases. Future studies will aim to understand the contribution of adult neurogenesis to the etiology and mechanisms of neurological diseases and disorders, and their treatments

Financial and Competing Interests Disclosure

The author has no relevant affiliations or financial involvement with any organization or entity with a financial interest in or financial conflict with the subject matter or materials discussed in the manuscript. This includes employment, consultancies, honoraria, stock ownership or options, expert testimony, grants or patents received or pending, or royalties.

No writing assistance was utilized in the production of this manuscript.

Acknowledgments

Reproduced from Expert Review of Neurotherapeutics 8(2), 311-320 (2008) with permission of Expert Reviews Ltd.

References

Papers of special note have been highlighted as:
• of interest
•• of considerable interest

[1] Santiago Ramon y Cajal. Degeneration and Regeneration of the Nervous System. Hafner, NY, USA (1928).

[2] Altman J, Das GD. Autoradiographic and histological evidence of postnatal hippocampal neurogenesis in rats. *J. Comp. Neurol.* 124(3), 319–336 (1965).

[3] Altman J. Autoradiographic and histological studies of postnatal neurogenesis. IV. Cell proliferation and migration in the anterior forebrain, with special reference to persisting neurogenesis in the olfactory bulb. *J. Comp. Neurol.* 137(4), 433–457 (1969).

[4] Taupin P, Gage FH. Adult neurogenesis and neural stem cells of the central nervous system in mammals. *J. Neurosci. Res*. 69(6), 745–749 (2002).

[5] Gross CG. Neurogenesis in the adult brain: death of a dogma. *Nat. Rev. Neurosci.* 1(1), 67–73 (2000).

[6] Eriksson PS, Perfilieva E, Bjork-Eriksson T et al. Neurogenesis in the adult human hippocampus. *Nat. Med.* 4(11), 1313–1317 (1998).

•• First report of adult neurogenesis in humans. Tissues originated from cancer patients treated with bromodeoxyuridine, as part of their therapy.

[7] Curtis MA, Kam M, Nannmark U et al. Human neuroblasts migrate to the olfactory bulb via a lateral ventricular extension. *Science* 315(5816), 1243–1249 (2007).

•• First report of migration of newly generated neuronal cells in the human subventricular zone (SVZ).

[8] Luskin MB. Restricted proliferation and migration of postnatally generated neurons derived from the forebrain subventricular zone. *Neuron* 11(1), 173–189 (1993).

[9] Lois C, Alvarez-Buylla A. Long-distance neuronal migration in the adult mammalian brain. *Science* 264(5162), 1145–1148 (1994).

[10] Cameron HA, Woolley CS, McEwen BS, Gould E. Differentiation of newly born neurons and glia in the dentate gyrus of the adult rat. *Neuroscience* 56(2), 337–344 (1993).

[11] Stanfield BB, Trice JE. Evidence that granule cells generated in the dentate gyrusof adult rats extend axonal projections. *Exp.Brain Res*. 72(2), 399–406 (1988).

[12] Markakis EA, Gage FH. Adult-generated neurons in the dentate gyrus send axonal projections to field CA3 and are surrounded by synaptic vesicles. *J. Comp. Neurol.* 406 (4), 449–460 (1999).

[13] van Praag H, Schinder AF, Christie BR et al. Functional neurogenesis in the adult hippocampus. *Nature* 415(6875), 1030–1034 (2002).

• First evidence of functional neurogenesis in the adult mammalian brain. The authors used retrovirus expressing green fluorescent protein to infect newly generated neuronal cells in the adult dentate gyrus (DG) of rodents.

[14] Carlen M, Cassidy RM, Brismar H et al. Functional integration of adult-born neurons. *Curr. Biol.* 12(7), 606–608 (2002).

[15] Belluzzi O, Benedusi M, Ackman J, LoTurco JJ. Electrophysiological differentiation of new neurons in the olfactory bulb. *J. Neurosci.* 23(32), 10411–10418 (2003).

[16] Toni N, Teng EM, Bushong EA et al. Synapse formation on neurons born in the adult hippocampus. *Nat. Neurosci.* 10(6), 727–734 (2007).

[17] Esposito MS, Piatti VC, Laplagne DA et al. Neuronal differentiation in the adult hippocampus recapitulates embryonic development. J. Neurosci. 25(44), 10074–10086 (2005). Erratum in: *J.Neurosci.* 25(49), table of contents (2005).

[18] Kempermann G, Kuhn HG, Gage FH. More hippocampal neurons in adult mice living in an enriched environment. *Nature* 386(6624), 493–495 (1997).

[19] Kornack DR, Rakic P. Continuation of neurogenesis in the hippocampus of the adult macaque monkey. *Proc. Natl Acad. Sci. USA* 96(10), 5768–5773 (1999).

[20] Cameron HA, McKay RD. Adult neurogenesis produces a large pool of new granule cells in the dentate gyrus. *J. Comp. Neurol.* 435(4), 406–417 (2001).

[21] Taupin, P. Adult neurogenesis in the mammalian central nervous system: functionality and potential clinical interest. *Med. Sci. Monit.* 11(7), RA247–RA252 (2005).

[22] Reynolds BA, Weiss S. Generation of neurons and astrocytes from isolated cells of the adult mammalian central nervous system. *Science* 255(5052), 1707–1710 (1992).
• First isolation and characterization of neural progenitor and stem cells from the adult mammalian brain. Neural progenitor and stem cells were isolated and characterized from the region including the SVZ.

[23] Gage FH, Coates PW, Palmer TD et al. Survival and differentiation of adult neuronal progenitor cells transplanted to the adult brain. *Proc. Natl Acad. Sci. USA* 92(25), 11879–11883 (1995).

[24] Doetsch F, Caille I, Lim DA et al. Subventricular zone astrocytes are neural stem cells in the adult mammalian brain. *Cell* 97(6), 703–716 (1999).

[25] Johansson CB, Momma S, Clarke DL et al. Identification of a neural stem cell in the adult mammalian central nervous system. *Cell* 96(1), 25–34 (1999).

[26] Seri B, Garcia-Verdugo JM, McEwen BS, Alvarez-Buylla A. Astrocytes give rise to new neurons in the adult mammalian hippocampus. *J. Neurosci.* 21(18), 7153–7160 (2001).

[27] Gage FH. Mammalian neural stem cells. *Science* 287(5457), 1433–1438 (2000).

[28] Chiasson BJ, Tropepe V, Morshead CM, van der Kooy D. Adult mammalian forebrain ependymal and subependymal cells demonstrate proliferative potential, but only subependymal cells have neural stem cell characteristics. *J. Neurosci.* 19(11), 4462–4471 (1999).

[29] Kornblum HI, Geschwind DH. Molecular markers in CNS stem cell research: hitting a moving target. *Nat. Rev. Neurosci.* 2(11), 843–846 (2001).

[30] Seaberg RM, van der Kooy D. Adult rodent neurogenic regions: the ventricular subependyma contains neural stem cells, but the dentate gyrus contains restricted progenitors. *J. Neurosci.* 22(5), 1784–1793 (2002).

[31] Suslov ON, Kukekov VG, Ignatova TN, Steindler DA. Neural stem cell heterogeneity demonstrated by molecular phenotyping of clonal neurospheres. *Proc. Natl Acad. Sci. USA* 99(22), 14506–14511 (2002).

[32] Bull ND, Bartlett PF. The adult mouse hippocampal progenitor is neurogenic but not a stem cell. *J. Neurosci.* 25(47), 10815–10821 (2005).

[33] Taupin P. Neurogenesis in the adult central nervous system. *C. R. Biol.* 329(7), 465–475 (2006).

[34] Jin K, Peel AL, Mao XO, Xie L, CottrellBA, Henshall DC, Greenberg DA. Increased hippocampal neurogenesis in Alzheimer's disease. *Proc. Natl Acad. Sci. USA* 101(1), 343–347 (2004).

[35] Jin K, Galvan V, Xie L et al. Enhanced neurogenesis in Alzheimer's disease transgenic (PDGF-APPSw,Ind) mice. *Proc. Natl Acad. Sci. USA* 101(36), 13363–13367 (2004).

[36] Feng R, Rampon C, Tang YP et al. Deficient neurogenesis in forebrain-specific presenilin-1 knockout mice is associated with reduced clearance of hippocampal memory traces. Neuron 32(5), 911–926 (2001). Erratum in: *Neuron* 33(2), 313 (2002).

[37] Wen PH, Shao X, Shao Z et al. Overexpression of wild type but not an FAD mutant presenilin-1 promotes neurogenesis in the hippocampus of adult mice. *Neurobiol. Dis.* 10(1), 8–19 (2002).

[38] Donovan MH, Yazdani U, Norris RD et al. Decreased adult hippocampal neurogenesis in the PDAPP mouse model of Alzheimer's disease. *J. Comp. Neurol.* 495(1), 70–83 (2006).

[39] Verret L, Jankowsky JL, Xu GM et al. Alzheimer's-type amyloidosis in transgenic mice impairs survival of newborn neurons derived from adult hippocampal neurogenesis. *J. Neurosci.* 27(25), 6771–6780 (2007).

[40] Zhang C, McNeil E, Dressler L, Siman R. Long-lasting impairment in hippocampal neurogenesis associated with amyloid deposition in a knock-in mouse model of familial Alzheimer's disease. *Exp. Neurol.* 204(1), 77–87 (2007).

[41] Caselli RJ, Beach TG, Yaari R, ReimanEM. Alzheimer's disease a century later. *J.Clin. Psychiatr.* 67(11), 1784–1800 (2006).

[42] St George-Hyslop PH, Petit A. Molecular biology and genetics of Alzheimer's disease. *C. R. Biol.* 328(2), 119–130 (2005).

[43] Dodart JC, Mathis C, Bales KR, PaulSM. Does my mouse have Alzheimer's disease? *Genes Brain Behav.* 1(3), 142–155 (2002).

[44] German DC, Eisch AJ. Mouse models of Alzheimer's disease: insight into treatment. *Rev. Neurosci.* 15(5), 353–369 (2004).

[45] Cameron HA, Gould E. Adult neurogenesis is regulated by adrenal steroids in the dentate gyrus. *Neuroscience* 61(2), 203–209 (1994).

[46] Gould E, Tanapat P, McEwen BS et al. Proliferation of granule cell precursors in the dentate gyrus of adult monkeys is diminished by stress. *Proc. Natl Acad. Sci. USA* 95(6), 3168–3171 (1998).

[47] Pham K, Nacher J, Hof PR, McEwen BS. Repeated restraint stress suppresses neurogenesis and induces biphasic PSA-NCAM expression in the adult rat dentate gyrus. *Eur J. Neurosci.* 17(4), 879–886 (2003).

[48] Reif A, Fritzen S, Finger M et al. Neural stem cell proliferation is decreased in schizophrenia, but not in depression. *Mol.Psychiatr.* 11(5), 514–522 (2006).

[49] Parent JM, Yu TW, Leibowitz RT et al. Dentate granule cell neurogenesis is increased by seizures and contributes to aberrant network reorganization in the adult rat hippocampus. *J. Neurosci.* 17(10), 3727–3738 (1997).

[50] Miller MW, Nowakowski RS. Use of bromodeoxyuridine-immunohistochemistry to examine the proliferation, migration and time of origin of cells in the central nervous system. *BrainRes.* 457(1), 44–52 (1988).

[51] Taupin P. BrdU immunohistochemistry for studying adult neurogenesis: paradigms, pitfalls, limitations, and validation. *BrainRes. Rev.* 53(1), 198–214 (2007).

[52] Tada E, Parent JM, Lowenstein DH, FikeJR. X-irradiation causes a prolonged reduction in cell proliferation in the dentate gyrus of adult rats. *Neuroscience* 99(1), 33–41 (2000).

[53] Parent JM, Tada E, Fike JR, LowensteinDH. Inhibition of dentate granule cell neurogenesis with brain irradiation does not prevent seizure-induced mossy fiber synaptic reorganization in the rat. *J. Neurosci.* 19(11), 4508–4519 (1999).

[54] Blumcke I, Schewe JC, Normann S et al. Increase of nestin-immunoreactive neural precursor cells in the dentate gyrus of pediatric patients with early-onset temporal lobe epilepsy. *Hippocampus* 11(3), 311–321 (2001).

[55] Fahrner A, Kann G, Flubacher A et al. Granule cell dispersion is not accompanied by enhanced neurogenesis in temporal lobe epilepsy patients. *Exp. Neurol.* 203(2), 320–332 (2007).

[56] Curtis MA, Penney EB, Pearson AG et al. Increased cell proliferation and neurogenesis in the adult human Huntington's disease brain. *Proc. Natl Acad. Sci. USA* 100(15), 9023–9027 (2003).

• Report showing the newborn neuronal cells from the SVZ migrate toward the site of degeneration.

[57] Lazic SE, Grote H, Armstrong RJ et al. Decreased hippocampal cell proliferation in R6/1 Huntington's mice. *Neuroreport* 15(5), 811–813 (2004).

[58] Tattersfield AS, Croon RJ, Liu YW et al. Neurogenesis in the striatum of the quinolinic acid lesion model of Huntington's disease. *Neuroscience* 127(2), 319–332 (2004).

[59] White JK, Auerbach W, Duyao MP et al. Huntingtin is required for neurogenesis and is not impaired by the Huntington's disease CAG expansion. *Nat Genet.* 17(4), 404–410 (1997).

[60] Zhao M, Momma S, Delfani K et al. Evidence for neurogenesis in the adult mammalian substantia nigra. *Proc. Natl Acad. Sci. USA* 100(13), 7925–7930 (2003).

[61] Frielingsdorf H, Schwarz K, Brundin P, Mohapel P. No evidence for new dopaminergic neurons in the adult mammalian substantia nigra. *Proc. Natl Acad. Sci. USA* 101(27), 10177–10182 (2004).

[62] Tande D, Hoglinger G, Debeir T et al. New striatal dopamine neurons in MPTP-treated macaques result from a phenotypic shift and not neurogenesis. *Brain* 129(Pt 5), 1194–1200 (2006).

[63] Liu BF, Gao EJ, Zeng XZ et al. Proliferation of neural precursors in the subventricular zone after chemical lesions of the nigrostriatal pathway in rat brain. *Brain Res.* 1106(1), 30–39 (2006).

[64] Hoglinger GU, Rizk P, Muriel MP et al. Dopamine depletion impairs precursor cell proliferation in Parkinson disease. *Nat. Neurosci*. 7(7), 726–735 (2004).

[65] Baker SA, Baker KA, Hagg T. Dopaminergic nigrostriatal projections regulate neural precursor proliferation in the adult mouse subventricular zone. *Eur. J. Neurosci*. 20(2), 575–579 (2004).

[66] Borta A, Hoglinger GU. Dopamine and adult neurogenesis. *J. Neurochem*. 100(3), 587–595 (2007).

[67] Liu J, Suzuki T, Seki T et al. Effects of repeated phencyclidine administration on adult hippocampal neurogenesis in the rat. *Synapse* 60(1), 56–68 (2006).

[68] Gould E, Tanapat P. Lesion-induced proliferation of neuronal progenitors in the dentate gyrus of the adult rat. *Neuroscience* 80(2), 427–436 (1997).

[69] Magavi SS, Leavitt BR, Macklis JD. Induction of neurogenesis in the neocortex of adult mice. *Nature* 405(6789), 951–955 (2000).

[70] Jin K, Xie L, Mao XO, Greenberg DA. Alzheimer's disease drugs promote neurogenesis. *Brain Res*. 1085(1), 183–188 (2006).

[71] Malberg JE, Eisch AJ, Nestler EJ, DumanRS. Chronic antidepressant treatment increases neurogenesis in adult rat hippocampus. *J. Neurosci*. 20(24), 9104–9110 (2000).

[72] Malberg JE, Duman RS. Cell proliferation in adult hippocampus is decreased by inescapable stress: reversal by fluoxetine treatment. *Neuropsychopharmacology* 28(9), 1562–1571 (2003).

[73] Perera TD, Coplan JD, Lisanby SH et al. Antidepressant-induced neurogenesis in the hippocampus of adult nonhuman primates. *J. Neurosci*. 27(18), 4894–4901 (2007).

[74] Banasr M, Soumier A, Hery M et al. Agomelatine, a new antidepressant, induces regional changes in hippocampal neurogenesis. *Biol. Psychiatr*. 59(11), 1087–1096 (2006).

[75] Santarelli L, Saxe M, Gross C et al. Requirement of hippocampal neurogenesis for the behavioral effects of antidepressants. *Science* 301(5634), 805–809 (2003).

[76] Jacobs BL, Praag H, Gage FH. Adult brain neurogenesis and psychiatry: a novel theory of depression. *Mol. Psychiatr*. 5(3), 262–269 (2000).

[77] Holick KA, Lee DC, Hen R, Dulawa SC. Behavioral effects of chronic fluoxetine in BALB/cJ mice do not require adult hippocampal neurogenesis or the serotonin 1A Receptor. *Neuropsychopharmacology* 33(2), 406–417 (2008).

[78] David DJ, Klemenhagen KC, Holick KA et al. Efficacy of the MCHR1 antagonist N-[3-(1-{[4-(3,4-difluorophenoxy)phenyl] methyl}(4-piperidyl))-4-methylphenyl]-2-methylpropanamide (SNAP 94847) in mouse models of anxiety and depression following acute and chronic administration is independent of hippocampal neurogenesis. *J. Pharmacol. Exp. Ther*. 321(1), 237–248 (2007).

[79] Freundlieb N, Francois C, Tande D et al. Dopaminergic substantia nigra neurons project topographically organized to the subventricular zone and stimulate precursor cell proliferation in aged primates. *J.Neurosci*. 26(8), 2321–2325 (2006).

[80] Arrieta JL, Artalejo FR. Methodology, results and quality of clinical trials of tacrine in the treatment of Alzheimer's disease: a systematic review of the literature. *Age Ageing* 27(2), 161–179 (1998).

[81] Wilkinson DG, Francis PT, Schwam E et al. Cholinesterase inhibitors used in the treatment of Alzheimer's disease: the relationship between pharmacological effects and clinical efficacy. *Drugs Aging* 21(7), 453–478 (2004).
[82] McShane R, Areosa Sastre A, Minakaran N. Memantine for dementia. *Cochrane Database Syst Rev*. 2, CD003154 (2006).
[83] Wong ML, Licinio J. Research and treatment approaches to depression. *Nat. Rev. Neurosci.* 2(5), 343–351 (2001).
[84] Brunello N, Mendlewicz J, Kasper S et al. The role of noradrenaline and selective noradrenaline reuptake inhibition in depression. *Eur. Neuropsychopharmacology* 12(5), 461–475 (2002).
[85] Hindmarch I. Expanding the horizons of depression: beyond the monoamine hypothesis. *Hum. Psychopharmacol.* 16(3), 203–218 (2001).
[86] Owens MJ. Selectivity of antidepressants: from the monoamine hypothesis of depression to the SSRI revolution and beyond. *J. Clin. Psychiatr.* 65(Suppl 4), 5–10 (2004).
[87] Gross C, Zhuang X, Stark K et al. Serotonin1A receptor acts during development to establish normal anxiety-like behaviour in the adult. *Nature* 416(6879), 396–400 (2002).
[88] Nowakowski RS, Hayes NL. Stem cells: the promises and pitfalls *Neuropsychopharmacology* 25(6), 799–804 (2001).
[89] Nowakowski RS, Hayes NL. New neurons: extraordinary evidence or extraordinary conclusion? *Science* 288(5467), 771 (2000).
[90] Gould E, Gross CG. Neurogenesis in adult mammals: some progress and problems. *J. Neurosci.* 22(3), 619–623 (2002).
[91] Rakic P. Adult neurogenesis in mammals: an identity crisis. *J. Neurosci.* 22(3), 614–618 (2002).
[92] Taupin P. Protocols for studying adult neurogenesis: insights and recent developments. *Regenerative Med.* 2(1), 51–62 (2007).
[93] Endo Y, Nishimura JI, Kobayashi S, Kimura F. Long-term glucocorticoid treatments decrease local cerebral blood flow in the rat hippocampus, in association with histological damage. *Neuroscience* 79(3), 745–752 (1997).
[94] Pont F, Collet A, Lallement G. Early and transient increase of rat hippocampal blood-brain barrier permeability to amino acids during kainic acid-induced seizures. *Neurosci. Lett.* 184(1), 52–54 (1995).
[95] Tillfors M, Furmark T, Marteinsdottir I et al. Cerebral blood flow in subjects with social phobia during stressful speaking tasks: a PET study. *Am. J. Psychiatr.* 158(8), 1220–1226 (2001).
[96] Oby E, Janigro D. The blood–brain barrier and epilepsy. *Epilepsia* 47(11), 1761–1774 (2006).
[97] Desai BS, Monahan AJ, Carvey PM, Hendey B. Blood–brain barrier pathology in Alzheimer's and Parkinson's disease: implications for drug therapy. *Cell Transplant.* 16(3), 285–299 (2007).

[98] Rehen SK, McConnell MJ, Kaushal D et al. Chromosomal variation in neurons of the developing and adult mammalian nervous system. *Proc. Natl Acad. Sci. USA* 98(23), 13361–13366 (2001).

[99] Rehen SK, Yung YC, McCreight MP et al. Constitutional aneuploidy in the normal human brain. *J. Neurosci.* 25(9), 2176–2180 (2005).

[100] Vincent I, Rosado M, Davies P. Mitotic mechanisms in Alzheimer's disease? *J. Cell Biol.* 132(3), 413–425 (1996).

[101] Yang Y, Geldmacher DS, Herrup K. DNA replication precedes neuronal cell death in Alzheimer's disease. *J. Neurosci.* 21(8), 2661–2668 (2001).

[102] Taupin P. Neurogenesis and Alzheimer's disease. Drug Target Insights 1, 1–4 (2006).

[103] Boekhoorn K, Joels M, Lucassen PJ. Increased proliferation reflects glial and vascular-associated changes, but not neurogenesis in the presenile Alzheimer hippocampus. *Neurobiol. Dis.* 24(1), 1–14 (2006).

[104] Watanabe Y, Gould E, McEwen BS. Stress induces atrophy of apical dendrites of hippocampal CA3 pyramidal neurons. *Brain Res.* 588(2), 341–345 (1992).

[105] Sheline YI, Wang PW, Gado MH et al. 1996. Hippocampal atrophy in recurrent major depression. *Proc. Natl Acad. Sci. USA* 93(9), 3908–3913 (1996).

[106] Campbell S, Marriott M, Nahmias C, MacQueen GM. Lower hippocampal volume in patients suffering from depression: a meta-analysis. *Am. J. Psychiatr.* 161(4), 598–607 (2004).

[107] Colla M, Kronenberg G, Deuschle M et al. Hippocampal volume reduction and HPA-system activity in major depression. *J. Psychiatr. Res.* 41(7), 553–560 (2007).

[108] Czeh B, Michaelis T, Watanabe T et al. Stress-induced changes in cerebral metabolites, hippocampal volume, and cell proliferation are prevented by antidepressant treatment with tianeptine. *Proc. Natl Acad. Sci. USA* 98(22), 12796–12801 (2001).

[109] Feldmann RE Jr, Sawa A, Seidler GH. Causality of stem cell based neurogenesis and depression – to be or not to be, is that the question? *J. Psychiatr. Res.* 41(9), 713–723 (2007).

[110] Taupin P. Neurogenesis and the effects of antidepressants. Drug Target Insights 1, 13–17 (2006).

[111] Paizanis E, Hamon M, Lanfumey L. Hippocampal neurogenesis, depressive disorders, and antidepressant therapy. *Neural Plast.* 73754 (2007).

[112] Brezun JM, Daszuta A. Depletion in serotonin decreases neurogenesis in the dentate gyrus and the subventricular zone of adult rats. *Neuroscience* 89(4), 999–1002 (1999).

[113] Axelson DA, Doraiswamy PM, McDonaldWM et al. Hypercortisolemia and hippocampal changes in depression. *Psychiatr. Res.* 47(2), 163–173 (1993).

[114] Inagaki M, Matsuoka Y, Sugahara Y et al. Hippocampal volume and first major depressive episode after cancer diagnosis in breast cancer survivors. *Am. J. Psychiatr.* 161(12), 2263–2270 (2004).

[115] Bielau H, Trubner K, Krell D et al. Volume deficits of subcortical nuclei in mood disorders A postmortem study. *Eur. Arch. Psychiatr. Clin. Neurosci.* 255(6), 401–412 (2005).

[116] Campbell S, Macqueen G. The role of the hippocampus in the pathophysiology of major depression. *J. Psychiatr. Neurosci.* 29(6), 417–426 (2004).

[117] Minghetti L. Role of inflammation in neurodegenerative diseases. *Curr. Opin. Neurol.* 18(3), 315–321 (2005).

[118] Zilka N, Ferencik M, Hulin I. Neuroinflammation in Alzheimer's disease: protector or promoter? *Bratisl. Lek. Listy* 107(9–10), 374–383 (2006).

[119] Monje ML, Toda H, Palmer TD. Inflammatory blockade restores adult hippocampal neurogenesis. *Science* 302(5651), 1760–1765 (2003).

[120] Ekdahl CT, Claasen JH, Bonde S et al. Inflammation is detrimental for neurogenesis in adult brain. *Proc. Natl Acad. Sci. USA* 100(23), 13632–13637 (2003).

[121] Stoll G, Jander S. The role of microglia and macrophages in the pathophysiology of the CNS. *Prog. Neurobiol.* 58(3), 233–247 (1999).

[122] Vallieres L, Campbell IL, Gage FH, Sawchenko PE. Reduced hippocampal neurogenesis in adult transgenic mice with chronic astrocytic production of interleukin-6. *J. Neurosci.* 22(2), 486–492 (2002).

[123] [Packer MA, Stasiv Y, Benraiss A et al. Nitric oxide negatively regulates mammalian adult neurogenesis. *Proc. Natl Acad. Sci. USA* 100(16), 9566–9571 (2003).

Chapter VII

Adult Neural Stem Cells: A Promising Candidate for Regenerative Therapy in the CNS

Abstract

Owing to the fact that neurodegenerative diseases, cerebral strokes and traumatic injuries to the central nervous system (CNS) produce neurological deficits that result from neuronal loss, cell therapy is a prominent area of investigation for the treatment of neurological diseases and injuries. Although various cell types have been considered and evaluated for therapy in the CNS, there is still no cure for these ailments, and new alternatives for cellular therapy must be explored. With the recent confirmation that neurogenesis occurs in the adult brain, and the isolation and characterization of neural progenitor and stem cells from the adult CNS, new avenues are being considered for cellular therapy and hold the promise to cure a broad range of CNS diseases and injuries. In this manuscript, we will review and discuss the merits and pitfalls of the main cell types considered for therapy in the CNS, and the advantages of adult-derived neural stem cells (NSCs) for regenerative therapy in the nervous system, over other cell types.

Introduction

Cellular therapy is the replacement of unhealthy or damaged cells or tissues by new ones. In the CNS, cell transplantation has been proposed to replace lost neurons in neurodegenerative diseases, and after injuries (Gage and Fisher, 1991). Because neurodegenerative disorders and brain injuries results in neuronal loss, two strategies have been devised for the restoration of the degenerated pathways: the supply of the missing neurotransmitters and/or the supply of neurotrophic factors to prevent degeneration. Over the past decades, various cell types, like fetal-derived neuronal cells, non-neuronal cells, embryonic and fetal stem cells have been considered and evaluated for their merits to restore the neuronal pathways and compensate for the neurological deficits, in various animal models of CNS diseases and injuries, and in clinical trials. The recent advances in somatic cell nuclear transfer (SCNT), and isolation and characterization of NSCs from the adult brain open new perspectives for cellular therapy in the CNS (Lanza *et al.*, 1999; Goh *et al.*, 2003). In this manuscript, we will review various cell types considered and evaluated for cellular therapy in the CNS, their merits, pitfalls, and controversies over their therapeutic use.

Fetal Neuronal Cells

Neuronal cells derived from fetal tissue have been extensively evaluated in transplantation studies, and demonstrated restoration of functional deficits in various animal models of CNS diseases (Gage *et al.*, 1984; Isacson *et al.*, 1984; Palfi *et al.*, 1998), providing the proof of principle for clinical trial investigations. Particularly, in rodents and non-human primates, fetal dopaminergic neurons from the mesencephalon survive, establish dopaminergic innervations and restore baseline dopamine synthesis, when transplanted into the denervated striatum (Mahalik *et al.*, 1985; Doucet *et al.*, 1989; Studer *et al.*, 1998; Bjorklund *et al.*, 2003), making fetal neuronal cell transplantation a promising strategy for the treatment of Parkinson's disease (PD). Clinical trials of fetal neuronal cell transplantation in PD have reported less conclusive results. In one study, mesencephalic tissue, derived form human embryos, resulted in some clinical benefits after transplantation in PD patients (Freed *et al.*, 2001), while another study reported no improvement after fetal nigral tissue transplantation (Olanow *et al.*, 2003). In patients with Huntington's disease, motor and cognitive improvements were reported 2 years after fetal neuronal cell transplantation (Bachoud-Levi *et al.*, 2000). However, these clinical improvements faded-off 4-6 years after the surgery (Bachoud-Levi *et al.*, 2006). Hence, the potential of fetal neuronal cell transplantation for cellular therapy remains to be further investigated and validated.

Besides the therapeutic merit, there are additional limitations to the use of fetal neuronal tissue for cellular therapy. The rate of survival of fetal cells transplanted in the adult brain is relatively low, requiring the isolation and purification of large quantities of cells for therapy, generally from several fetuses (Dunnett and Bjorklund, 1999). Since autologous tissue cannot be used as donor for transplantation, immunosuppressant therapy is needed to avoid graft rejection, and there are also ethical and political concerns associated with the use of human fetuses for research and therapy (McLaren, 2001). To overcome these limitations, sources of tissues from other species, in xenotransplantation studies, have been evaluated (Larsson *et al.*, 2000; Barker *et al.*, 2000). Pigs are the animals being the most considered for such studies, as they generate large litters (the number of dopaminergic neuronal cells in the developing pig brain is estimated to be 200,000, which makes it a source of choice for fetal neuronal cell therapy for PD) (Weiss, 1998). There are however limitations to xenogeneic transplantation for cellular therapy, like the risk of zoonotic infection, of immune-rejection, as well as ethical issues (Isacson and Breakefield, 1997; George, 2006).

Altogether these data show that despite some success and potential for the treatment of neurodegenerative diseases, allografted fetal neuronal human tissue and xenografted neural tissue for cellular therapy remain limited by technical, safety, and ethical issues.

Paracrine Systems

Various non-neuronal cell types have been investigated for their ability to locally deliver neurotransmitters and/or trophic factors in the CNS after grafting, functioning as "minipumps", to compensate for the neuronal loss or rescue degenerating nerve cells, to promote functional recovery. Among them, endocrine cells secreting neurotransmitters and trophic factors, and genetically modified cells engineered to secrete neurotransmitters or trophic factors have been proposed, as sources of transplants. Cells derived from the sympathoadrenal (SA) lineage, like chromaffin cells - the neuroendocrine cells of the adrenal medulla- and sympathetic neurons, mostly release noradrenaline, though some of them are able to produce and release dopamine. SA cells express also dopaminotrophic factors, like glial-derived neurotrophic factor and transforming growth factor-β, which protect dopaminergic neurons from degeneration (Unsicker *et al.*, 1997). SA cells have been proposed for the treatment of PD, and extensively investigated in transplantation studies in animal models of PD, and proposed for cell-replacement therapy for PD. One of the advantages of SA transplantation is the ability to isolate the cells from the patient himself, allowing autologous transplantation, thereby obviating the need for immunosuppressant therapy (Fernandez-Espejo *et al.*, 2005). Adrenal medulla cells and chromaffin cells transplanted into the denervated striatum exert beneficial effects in animal models of PD and in patients (Backlund *et al.*, 1985; Madrazo *et al.*, 1987; Fernandez-Espejo *et al.*, 2005). Since the proportion of dopaminergic cells in SA tissue transplanted is very low -only 1% of the entire adrenal chromaffin cell population releases dopamine-, it is proposed that the beneficial activity of transplanted adrenal cells on PD symptoms results in its neurotrophic effect, rather the release of dopamine (Fernandez-Espejo *et al.*, 2005; Brown and Dunnet; 1989). Further, the survival of adrenal medulla grafts is low in animal models of PD and extremely low after grafting in PD patients (Backlund *et al.*, 1985; Madrazo *et al.*, 1987; Brown and Dunnet; 1989). Hence, despite the ability to isolate and purify adrenal cells in extremely high quantities and to perform autologous transplantation, this approach is no longer pursued clinically. Reports of transplanted sympathetic neurons, either freshly isolated or after culture, in animal models of PD show that while freshly isolated sympathetic neurons elicit poor survival and exerted limited beneficial activity when transplanted intrastriatally, cultured sympathetic neurons elicit a more robust beneficial effect (Stenevi *et al.*, 1976), leading to clinical trials of cultured autologous sympathetic neurons in PD patients (Nakao *et al.*, 2001). Results from clinical trials showed that cultured sympathetic neuron grafts induce a partial symptomatic relief in PD patients. However, as for chromaffin cells, the main limitation for the use of sympathetic neurons for therapy is their poor survival (Stenevi *et al.*, 1976; Nakao *et al.*, 2001).

Non-neuronal cell types of other origins have been proposed for the treatment of PD in xenotransplantation studies. Among them, bovine adrenal medullary chromaffin cells and PC12 cells, a rat-derived pheochromocytoma cell line secreting L-dopa and dopamine (Subramanian *et al.*, 1997; Lindner *et al.*, 1998). Prior to transplantation, these cells are encapsulated within polymer membranes, to avoid immune rejection -by immuno-isolating them from the recipient immune system- and tumor formation, while allowing functional efficacy (Li *et al.*, 1999; Gray, 2001). Xenografts of PC12 cells can survive for up to 6.5

months in non-immunosuppressed monkeys when immuno-isolated via polymer encapsulation, and continue to secrete high levels of levodopa and dopamine, and induce recovery of motor function in parkinsonian nonhuman primates (Kordower *et al.*, 1995). While this strategy appear to overcome some of the limitations associated with fetal-derived neuronal tissue and xenotransplantation, such as technical, safety and ethical issues, fundamental issues remain to be addressed, particularly with regard to the duration and consistency of cell viability and device output.

Genetically engineered non-neuronal cells, such as fibroblasts and astrocytic cell lines, to express neuromediators or trophic factors has also been proposed for the treatment of CNS diseases and injuries (Palmer *et al.*, 1991; Fisher *et al.*, 1991; Winkler *et al.*, 1995; Tornatore *et al.*, 1996; Tuszynski *et al.*, 1996). These cells may also constitutively express factors with neurotrophic potential, further enhancing their regenerative potential. Recently, fibroblasts genetically engineered to express nerve growth factor transplanted in patients with Alzheimer's disease, have been shown to improve the patients' abilities to recover (Tuszynski *et al.*, 2005), validating such strategy for cellular therapy. Fibroblasts can be easily isolated from the own patients, allowing autologous transplantation, thereby providing an interesting model for cellular therapy. However, the main concern over the use of genetically engineered cells for cellular therapy resides over the long-term expression of the transgene and the need to develop vectors allowing sustained expression of the transgene (Palmer *et al.*, 1991; Wei *et al.*, 1999; Pizzo *et al.*, 2004).

Altogether, the use of paracrine systems for cellular therapy in the CNS offers a wide range of strategy and the possibility of autologous transplantation, but their main limitations are the low survival of the grafted cells, the long-term release and expression of the transgene, by encapsulated or genetically engineered cells, and their inability to restore neuronal circuitries and a controlled synaptic release of transmitter.

Stem Cells

Embryonic Stem Cells

Embryonic stem cells (ESCs) are the self-renewing, pluripotent cells that can generate all the cell types of the body, and therefore carry the hope to cure a broad range of diseases, particularly for the CNS (Wobus and Boheler, 2005). They have the ability to remain undifferentiated and to proliferate indefinitely *in vitro,* while maintaining the potential to differentiate into derivatives of all three embryonic germ layers, the ectoderm, mesoderm and endoderm. ESCs are derived from the inner cellular mass (ICM) of blastocysts (Figure 1), and have been isolated and cultured from human blastocysts, a milestone for cellular therapy (Thomson *et al.*, 1998). Protocols have been devised to differentiate ESCs to the neuronal pathways (Brustle *et al.*, 1999; Zhang *et al.*, 2001; Tropepe *et al.*, 2001; Nakayama *et al.*, 2004; Kato *et al.*, 2006), and ESCs have been successfully grafted in animal models of neurodegenerative diseases. Particularly, neurons with dopaminergic phenotype have been generated in culture from mouse and primate ESCs, transplanted in dopamine-depleted striatum and shown to improve deficits in animal models of PD (Kawasaki *et al.*, 2000; Lee *et al.*, 2000; Perrier *et al.*, 2004; Bjorklund *et al.*, 2002; Kim *et al.*, 2002). Recently, Keirstad *et al.* (2005) reported the successful transplantation and integration of human ESCs in an experimental model of spinal cord injury. Reconstitution of the myelin and functional recovery were reported, suggesting that such strategy could be applied to human injuries.

There are however technical, ethical and political limitations to overcome to the use of ESCs for cellular therapy (Taupin, 2006a). To maintain their stem cell properties, and thus therapeutic potential, original protocols devised to culture mouse ESCs required the cells to be cultured on fibroblast feeder layer, derived from murine embryos; the murine embryonic feeder layer providing the cellular and molecular cues for the maintenance of the ESC properties *in vitro* (Evans and Kaufman, 1981). These conditions were also used to derive the first lines of human ESCs (Thomson *et al.*, 1998), a limiting factor for the use of these cell lines for therapy. Recently, it was reported these established human ESC lines were contaminated with N-glycolyl-neuraminic (Neu5Gc) acid residues, likely originating from the mouse feeder layer used to derive the cell lines (Martin *et al.*, 2005). Neu5Gc is a sugar present on the surface of most mammal and rodent cells, but not in humans (Muchmore *et al.*, 1998). Human cells can capture Neu5Gc -from dietary sources for example (Tangvoranuntakul *et al.*, 2003; Bardor *et al.*, 2005)-, as such most humans have developed circulating antibodies against it (Merrick *et al.*, 1978). Though recent studies revealed that the incorporation of Neu5Gc by the established human ESCs is reversible (Heiskanen *et al.*, 2007), it could result potentially in the rejection of the human ESCs upon grafting, mandating for the generation of new cell lines devoid of animal contaminants (Martin *et al.*, 2005). The combination of fibroblast growth factor 2 and antagonists of bone morphogenetic protein has been reported to maintain human ESCs pluripotentiality, in the absence of mouse feeder cells (Xu *et al.*, 2005), and new human ESC lines have been derived free of feeder layer, in defined medium (Ludwig *et al.*, 2006), providing a source of tissue for cellular therapy. Though ESCs propagate indefinitely in culture, the maintenance of their karyotypes overtime has been the source of debates and controversies. On the one hand reports suggest that human ESCs may not maintain their normal karyotypes, while others have confirmed that some established cells line remain stable overtime (Draper *et al.*, 2004; Buzzard *et al.*, 2004; Mitalipova *et al.*, 2005; Brimble *et al.*, 2001). As difference is handling procedures may account for these discrepancies and karyotype variations have been linked to cell transformation, the maintenance of the established cell lines must be checked overtime. ESCs have the potential to form tumors upon grafting (Thomson *et al.*, 1998; Wakitani *et al.*, 2003). The formation of teratomes is associated with the undifferentiated state of the ESCs. It is proposed to pre-differentiate the ESCs *in vitro* to the desired lineage, and to remove the cells that have not differentiated prior to grafting. There are three strategies that could be considered to reduce the risk of tumor formation: 1) devising condition that would yield to a fully differentiated culture, 2) genetically engineered ESCs to introduce drug-resistant genes under the control of a lineage-specific promoter, and 3) purifying, by positive or negative selection, the differentiated cells, with the use of cell surface markers and fluorescence activated cell sorting for example. The derivation of ESCs from embryos lead to the establishment of allogenic cell lines for cellular therapy that would require to establish a bank of ESC lines, and matching the donor and the recipient for histocompatibility, to reduce the risk of tissue rejection, or to administer immune-suppressive drugs, like cyclosporine, upon transplantation to the patient. The generation of ESC lines for cellular therapy requires the destruction of blastocysts. The use of human fetuses, embryos and blastocysts for research and therapy is strictly regulated (McLaren, 2001). To overcome such limitations, researchers have attempted to derive ESCs without destroying blastocysts. Recently, Chung *et al.* (2006) reported the isolation of single cells from mouse blastomeres. The isolated cells behaved like ESCs *in vitro*, while the embryos (composed of 7 cells) went on pursuing their normal development, after implantation in pseudo-gestant female. Though similar to pre-implantation genetic diagnosis used in fertility clinics, the use of such strategy to derive human ESC lines remain to be demonstrated and would not be without ethical limitations.

Altogether these data show that though ESCs hold the promise to cure a broad range of diseases, particularly for the CNS. However, their use in therapy faces many challenges and limitations.

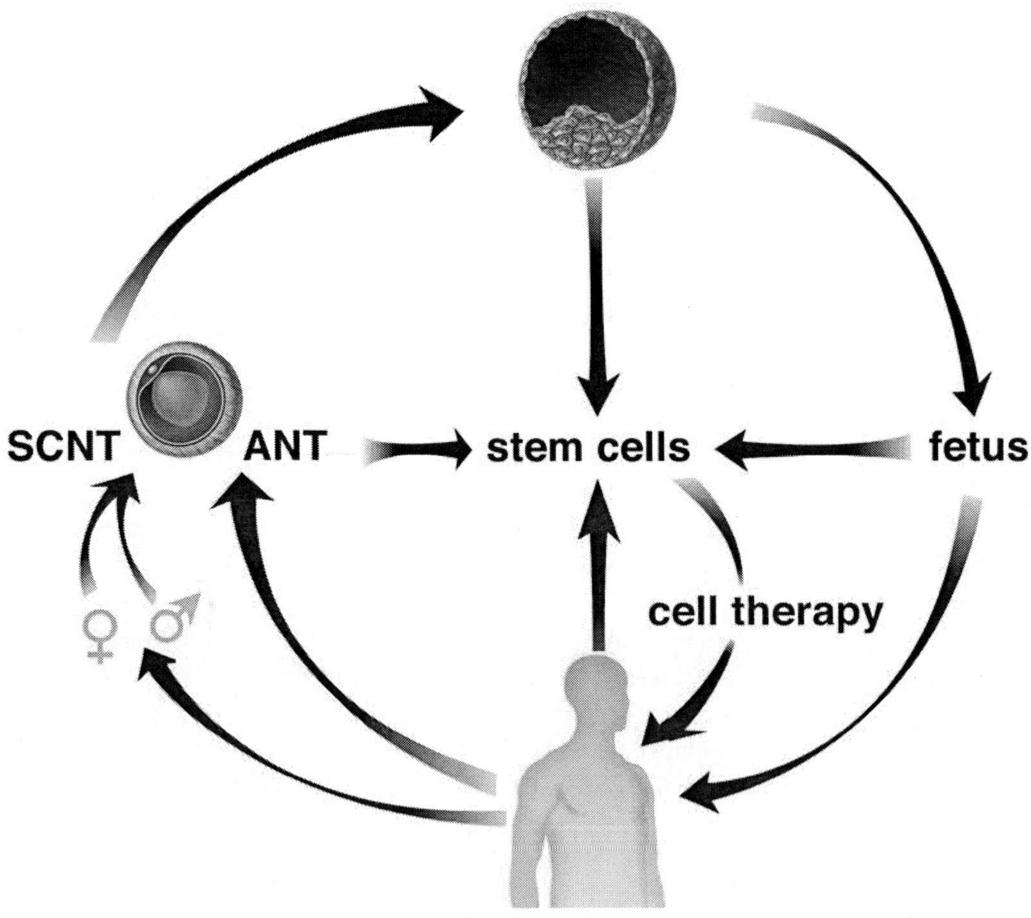

Figure 1. Source of stem cells for cellular therapy. Stem cells can be isolated at various stage of development. Stem cells can be derived from embryos (blastocysts), the embryonic stem cells (ESCs), from fetuses and from adult tissues. Isogenic ESCs lines can be derived from cloned embryos by somatic cell nuclear transfer (SCNT). Strategies have been reported to derive ESCs lines without the destruction of embryos from blastomeres or by a procedure derived from nuclear transfer; altered nuclear transfer (ANT).

Cloned Embryonic Stem Cells

Therapeutic cloning aims at generating cell lines, tissues, and organs that would have the patient own genetic make up, and thus not be rejected. With the recent advance in SCNT, there is the potential to generate cloned stem cells, tissues, organs and individuals by transferring the nucleus of somatic cells harvested from donors into enucleated oocytes (Campbell *et al.*, 1996; Rhind *et al.*, 2003) (Figure 1). SCNT has been successfully applied to clone various animals, such as sheep, mice, pigs, rabbits, cats and dogs (Campbell *et al.*, 1996; Wakayama *et al.*, 2000; Polejaeva *et al.*, 2000; Chesne *et al.*, 2002; Shin *et al.*, 2002; Rhind *et al.*, 2003; Lee *et al.*, 2005). Further reports showing that ESCs derived by nuclear transfer can be used to correct immunodeficiency in mice (Rideout *et al.*, 2002) or that c-kit+ cells isolated from cloning-derived fetuses could engraft in infracted tissue, differentiate into cardiomyocytes, repopulate a substantial fraction of the scar and improve ventricular hemodynamics (Lanza *et al.*, 2004), have validated the potential of cloned ESC lines, and thus SCNT, for cellular therapy.

There are however strong limitations to overcome for the use of the cloned cell lines for therapy, particularly unknown regarding the viability and developmental potential of the generated cell lines and tissues (Shiels *et al.*, 1999; Tian *et al.*, 2000; Wilmut, 2003). There are also ethical concerns over therapeutic cloning, and the generation of human cloned ESCs raises the same ethical issue than ESCs isolated from donated blastocysts. To overcome such limitation, researchers have attempted to derive cloned ESCs without destroying blastocysts. Recently, Meissner and Jaenisch (2005) reported the isolation of cloned mouse ESCs by a procedure derived from SCNT, called altered nuclear transfer (ANT) (Meissner and Jaenisch, 2006). ANT is based on the inactivation of a gene crucial for trophectoderm development, such as the gene *Cdx2* that encodes the earliest-known trophectoderm-specific transcription factor. The inactivation of *Cdx2* eliminates the potential to form the fetal-maternal interface. The resultant blastocysts are unable to implant into the uterus and to continue their development, but their ICMs are spared, from which ESCs could be derived (Hurlbut, 2005). The investigators reported a protocol for conditionally inactivating *Cdx2*, using a lentil virus. The cloned blastocysts were morphologically abnormal, lacked functional trophoblast, and failed to implant into the uterus. Yet, the eggs divided and grew enough, so that ESCs could be derived, after re-establishment of *Cdx2* expression (Tangvoranuntakul *et al.*, 2003). Though this technique yields to the derivation of ESC lines, ANT remains the subject to debates and controversies, and does not resolve the ethical issue associated with the destruction of blastocysts (Melton *et al.*, 2004). Hence, it is unlikely that such procedure would be applied to derive human ESCs. Further the consequences of the inactivation of *Cdx2* on the normal development remain unknown, and pose additional ethical issues.

Though these data show that SCNT has the potential to generate ESCs that have the genetic make-up of the individuals, the ethical issues surrounding the destruction of blastocysts remain a major limitation for the therapeutic application of SCNT-derived ESCs. The generation of isogenic ESCs by SCNT from human remains also to be established (Kennedy, 2006).

Neural Stem Cells

SCs are the self-renewing, multipotent cells that generate the main phenotypes of the nervous system, neuronal, astrocytic, and oligodendrocytic. Because of their potential to generate the different cell types of the CNS, NSCs hold the promise to cure a broad range of CNS diseases and injuries (Gage, 2000). Neural progenitor and stem cells can be isolated from fetal tissues (Figure 1). However, fetal-derived neural progenitor and stem cells carry the same ethical and political limitations than fetal primary neuron cultures and ESCs (McLaren, 2001). The recent confirmation that neurogenesis occurs in the adult brain, and the isolation of neural progenitor and stem cells from adult tissues offers new sources of tissues for cellular therapy in the CNS (Reynolds and Weiss, 1992; Gross, 2000; Taupin and Gage, 2002), without the ethical and political hurdles associated with the use of fetal tissues (Figure 1). Neural progenitor and stem cells have been isolated and grafted in various animal models of CNS diseases and injuries, validating their use for therapy (Gage *et al.*, 1995; Shihabuddin *et al.*, 2000; Burnstein *et al.*, 2004; Iwanami *et al.*, 2005; Cummings *et al.*, 2005; Murrell *et al.*, 2005; Reynolds and Rietze, 2005; Taupin, 2005; Uchida *et al.*, 2000). Although further experiments are needed, grafted neural progenitor and stem cells show functional integration and promote functional recovery, and could overcome the problem of the low survival rate of fetal neurons. Furthermore, NSCs offer several advantages over other cell types for cellular therapy, such as ESCs that have the risk to form tumors upon grafting, and over the so-called "paracrine systems", as they permit the rewiring of the CNS. The ability to isolate and culture neural progenitor and stem cells from adult tissues open also the opportunity to perform autologous transplantation, in which neural progenitor and stem cells would be isolated from un undamaged areas of the CNS, expanded *in vitro*, and grated back -with or without prior differentiation- to repair the CNS, thereby, obviating the need of donor-recipient matching, or the use of anti-rejection drugs, conditions that would favor successful graft integration, survival, and recovery. In support to this contention, neural progenitor and stem cells can be isolated and expanded *in vitro* from various areas of the adult CNS, including the spinal cord, providing various potential sites to harvest the cells for transplantation. Experimental studies in rodents have shown that neural progenitor and stem cells isolated from the adult spinal cord and grafted in the spinal cord, differentiate into glial cells, whereas when transplanted in the hippocampus, differentiate into neuronal and glial cells (Shihabuddin *et al.*, 2000), as hippocampal-derived neural progenitor and stem cells (Gage *et al.*, 1995). These data suggest that neural progenitor and stem cells isolated from various areas have the potential to engraft in heterotypic areas, validating autologous transplantation of adult-derived NSCs as a strategy for cellular therapy. There are however limitations to the isolation of adult NSCs for autologous transplantation, such as the expected serious consequences of an invasive surgical procedure and the possible permanent damage to the donor. Recently, neural progenitor and stem cells have been isolated and characterized from the adult olfactory neuroepithelium, providing a source of autologous of stem cells for cellular therapy in the CNS that can be easily established without invasive surgery and expected serious consequences to the donor (Murrell *et al.*, 2005).

There are however limitations to the use of NSCs for cellular therapy. Currently established protocols to isolated NSCs yield to heterogeneous populations of neural progenitor and stem cells (Taupin and Gage, 2002; Reynolds and Rietze, 2005; Taupin, 2005). Recent studies have tackled these issues, and homogenous population of neural progenitor/stem cells have been isolated and characterized *in vitro*, using cell surface markers (Uchida *et al.*, 2000; Capela and Temple, 2002; Nagato *et al.*, 2005), by promoter-targeted selection (Sawamoto *et al.*, 2001), and by "side-population" analysis (Kim and Morshead, 2003). There are other limitations to the use of neural progenitor and stem cells for cellular therapy, such as the uncertain potential to differentiate to specific neuronal phenotypes, to establish the right connections, and the potential to form tumor upon grafting (Taupin, 2006b).

Altogether, these data show that neural progenitor and stem cells provide a promising model for cellular therapy, particularly adult NSCs that are not associated with ethical and political concern. There is also the potential for adult NSCs to perform autologous transplantation, however such strategy remains to be validated. Adult neural progenitor and stem cells can also be isolated from human *post-mortem* tissues (Schwartz *et al.*, 2003), potentially allowing the generation of neural progenitor and stem cells from multiple sources for cellular therapy, as alternative strategies, as well as to establish NSC banks. Future studies will aim to further evaluate the potential of adult-derived NSCs for cellular therapy.

Conclusion

Various sources of tissues are considered and evaluated for cellular therapy in the CNS. Among them, ESCs holds the promise to treat a broad range of CNS diseases and injuries. However, potential tumorigenicity, immunogenicity, ethical and political concerns represent major risks and hurdles that limit the use of ESCs for therapy, and will need to be addressed. The confirmation that neurogenesis occurs in the adult brain and NSCs reside in the adult CNS offers new avenues for cellular therapy. Adult derived-NSCs are not associated with ethical and political concerns and therefore represent a promising candidate for regenerative therapy in the nervous system. Further, because NSCs reside in the adult CNS, the stimulation of endogenous neural progenitor and stem cells in the adult brain may provide an alternative strategy for cellular therapy in the CNS, using adult NSCs.

Acknowledgments

Reproduced with permission from International Journal of Integrative Biology: Taupin P. Adult neural stem cells: a promising candidate for regenerative therapy in the CNS. International Journal of Integrative Biology (2008) 2(2):85-94. Copyrights: International Journal of Integrative Biology.

References

Bachoud-Levi A.C., Remy P. *et al.* (2000). Motor and cognitive improvements in patients with Huntington's disease after neural transplantation. *Lancet*. 356(9246): 1975-1979.

Bachoud-Levi A.C., Gaura V. *et al.* (2006). Effect of fetal neural transplants in patients with Huntington's disease 6 years after surgery: a long-term follow-up study. *Lancet Neurol.* 5(4): 303-309.

Backlund E.O., Granberg P.O. *et al.* (1985). Transplantation of adrenal medullary tissue to striatum in parkinsonism. First clinical trials. *J. Neurosurg*. 62(2): 169-173.

Bardor M., Nguyen D.H. *et al.* (2005). Mechanism of uptake and incorporation of the non-human sialic acid N-glycolylneuraminic acid into human cells. *J. Biol. Chem*. 280(6): 4228-4237.

Barker R.A., Kendall A.L., Widner H. (2000). Neural tissue xenotransplantation: what is needed prior to clinical trials in Parkinson's disease? Neural Tissue Xenografting Project. *Cell Transplant*. 9(2): 235-246.

Bjorklund L.M., Sanchez-Pernaute R. *et al.* (2002). Embryonic stem cells develop into functional dopaminergic neurons after transplantation in a Parkinson rat model. *Proc. Natl. Acad. Sci. USA* 99(4): 2344-2349.

Bjorklund A., Dunnett S.B. *et al.* (2003). Neural transplantation for the treatment of Parkinson's disease. *Lancet Neurol*. 2(7): 437-445.

Brimble S.N., Zeng X. *et al.* (2004). Karyotypic stability, genotyping, differentiation, feeder-free maintenance, and gene expression sampling in three human embryonic stem cell lines derived prior to August 9, 2001. *Stem Cells Dev*. 13(6): 585-597.

Brustle O., Jones K.N. *et al.* (1999). Embryonic stem cell-derived glial precursors: a source of myelinating transplants. *Science* 285(5428): 754-756.

Brown V.J., Dunnett S.B. (1989) Comparison of adrenal and foetal nigral grafts on drug-induced rotation in rats with 6-OHDA lesions. *Exp. Brain Res*. 78(1): 214-218.

Burnstein R.M., Foltynie T. *et al.* (2004). Differentiation and migration of long term expanded human neural progenitors in a partial lesion model of Parkinson's disease. *Int. J. Biochem. Cell Biol*. 36(4): 702-713.

Buzzard J.J., Gough N.M. *et al.* (2004). Karyotype of human ES cells during extended culture. *Nat. Biotechnol*. 22(4): 381-382.

Campbell K.H., McWhir J. *et al.* (1996). Sheep cloned by nuclear transfer from a cultured cell line. *Nature 380*(6569): 64-66.

Capela A., Temple S. (2002). LeX/ssea-1 is expressed by adult mouse CNS stem cells, identifying them as nonependymal. *Neuron* 35(5): 865-875.

Chesne P., Adenot P.G. *et al.* (2002). Cloned rabbits produced by nuclear transfer from adult somatic cells. *Nat. Biotechnol*. 20(4): 366-369.

Chung Y., Klimanskaya I. *et al.* (2006). Embryonic and extraembryonic stem cell lines derived from single mouse blastomeres. *Nature* 439(7073): 216-219.

Cummings B.J., Uchida N. *et al.* (2005). Human neural stem cells differentiate and promote locomotor recovery in spinal cord-injured mice. *Proc. Natl. Acad. Sci. USA* 102(39): 14069-14074.

Doucet G., Murata Y. *et al.* (1989) Host afferents into intrastriatal transplants of fetal ventral mesencephalon. *Exp. Neurol.* 106(1): 1-19.

Draper J.S., Smith K. *et al.* (2004). Recurrent gain of chromosomes 17q and 12 in cultured human embryonic stem cells. *Nat. Biotechnol.* 22(1): 53-54.

Dunnett S.B., Bjorklund A. (1999). Prospects for new restorative and neuroprotective treatments in Parkinson's disease. *Nature* 399(6738 Suppl): A32-39.

Evans M.J., Kaufman M.H. (1981). Establishment in culture of pluripotential cells from mouse embryos. *Nature* 292(5819): 154-156.

Fernandez-Espejo E., Armengol J.A. *et al.* (2005). Cells of the sympathoadrenal lineage: biological properties as donor tissue for cell-replacement therapies for Parkinson's disease. *Brain Res. Brain Res. Rev.* 49(2): 343-354.

Fisher L.J., Jinnah H.A. *et al.* (1991). Survival and function of intrastriatally grafted primary fibroblasts genetically modified to produce L-dopa. *Neuron* 6(3): 371-380.

Freed C.R., Greene P.E. *et al.*, (2001). Transplantation of embryonic dopamine neurons for severe Parkinson's disease. *N. Engl. J. Med.* 344(10): 710-719.

Gage F.H. (2000). Mammalian neural stem cells. *Science* 287(5457): 1433-1438.

Gage F.H., Bjorklund A. et al., (1984). Intrahippocampal septal grafts ameliorate learning impairments in aged rats. *Science* 225(4661): 533-536.

Gage F.H., Coates P.W. *et al.* (1995). Survival and differentiation of adult neuronal progenitor cells transplanted to the adult brain. *Proc. Natl. Acad. Sci. USA* 92(25): 11879-11883.

Gage F.H., Fisher L.J. (1991). Intracerebral grafting: a tool for the neurobiologist. *Neuron* 6(1): 1-12.

George J.F. (2006) Xenotransplantation: an ethical dilemma. *Curr. Opin. Cardiol.* 21(2): 138-141.

Goh E.L., Ma D. *et al.* (2003). Adult neural stem cells and repair of the adult central nervous system. *J. Hematother. Stem Cell Res.* 12(6): 671-679.

Gray D.W. (2001). An overview of the immune system with specific reference to membrane encapsulation and islet transplantation. *Ann. N Y Acad. Sci.*944: 226-239.

Gross C.G. (2000). Neurogenesis in the adult brain: death of a dogma. *Nat. Rev. Neurosci.* 1(1): 67-73.

Heiskanen A., Satomaa T. *et al.* (2007). N-glycolylneuraminic acid xenoantigen contamination of human embryonic and mesenchymal stem cells is substantially reversible. *Stem Cells* 25(1): 197-202.

Hurlbut W.B. (2005). Altered nuclear transfer as a morally acceptable means for the procurement of human embryonic stem cells. *Perspect. Biol. Med.* 48(2): 211-228.

Isacson O., Brundin P. *et al.* (1984). Functional neuronal replacement by grafted striatal neurones in the ibotenic acid-lesioned rat striatum. *Nature* 311(5985): 458-460.

Isacson O., Breakefield X.O. (1997). Benefits and risks of hosting animal cells in the human brain. *Nat. Med.* 3(9): 964-969.

Iwanami A., Kaneko S. *et al.* (2005). Transplantation of human neural stem cells for spinal cord injury in primates. *J. Neurosci. Res.* 80(2): 182-190.

Kato T., Heike T. *et al.* (2006). A neurosphere-derived factor, cystatin C, supports differentiation of ES cells into neural stem cells. *Proc. Natl. Acad. Sci. USA* 103(15): 6019-6024.

Kawasaki H., Mizuseki K. *et al.* (2000). Induction of midbrain dopaminergic neurons from ES cells by stromal cell-derived inducing activity. *Neuron* 28(1): 31-40.

Keirstead H.S., Nistor G. *et al.* (2005). Human embryonic stem cell-derived oligodendrocyte progenitor cell transplants remyelinate and restore locomotion after spinal cord injury. *J. Neurosci.* 25(19):4694-4705.

Kennedy D. (2006). Editorial retraction. *Science* 311(5759): 335.

Kim J.H., Auerbach J.M. *et al.* (2002). Dopamine neurons derived from embryonic stem cells function in an animal model of Parkinson's disease. *Nature* 418(6893): 50-56.

Kim M., Morshead C.M. (2003). Distinct populations of forebrain neural stem and progenitor cells can be isolated using side-population analysis. *J. Neurosci.* 23(33): 10703-10709.

Kordower J.H., Liu Y.T. *et al.* (1995). Encapsulated PC12 cell transplants into hemiparkinsonian monkeys: a behavioral, neuroanatomical, and neurochemical analysis. *Cell. Transplant.* 4(2): 155-171.

Lanza R.P., Cibelli J.B., West M.D. (1999). Prospects for the use of nuclear transfer in human transplantation. *Nat. Biotechnol.* 17(12): 1171-1174.

Lanza R., Moore M.A. *et al.* (2004). Regeneration of the infarcted heart with stem cells derived by nuclear transplantation. *Circ. Res.* 94(6): 820-827.

Larsson L.C., Czech K.A. *et al.* (2000). Intrastriatal ventral mesencephalic xenografts of porcine tissue in rats: immune responses and functional effects. *Cell Transplant.* 9(2): 261-272.

Lee B.C., Kim M.K. *et al.* (2005). Dogs cloned from adult somatic cells. *Nature* 436(7051): 641. Erratum in: (2005). *Nature* 436(7054): 1102.

Lee S.H., Lumelsky N. *et al.* (2000). Efficient generation of midbrain and hindbrain neurons from mouse embryonic stem cells. *Nat. Biotechnol.* 18(6): 675-679.

Li R.H., Williams S. *et al.* (1999). Dose control with cell lines used for encapsulated cell therapy. *Tissue Eng.* 5(5): 453-466.

Lindner M.D., Emerich D.F. (1998). Therapeutic potential of a polymer-encapsulated L-DOPA and dopamine-producing cell line in rodent and primate models of Parkinson's disease. *Cell. Transplant.* 7(2): 165-174.

Ludwig T.E., Levenstein M.E. *et al.* (2006). Derivation of human embryonic stem cells in defined conditions. *Nat. Biotechnol.* 24(2): 185-187.

Madrazo I., Drucker-Colin R. *et al.* (1987) Open microsurgical autograft of adrenal medulla to the right caudate nucleus in two patients with intractable Parkinson's disease. *N. Engl. J. Med.* 316(14): 831-834.

Mahalik T.J., Finger T.E. *et al.* (1985). Substantia nigra transplants into denervated striatum of the rat: ultrastructure of graft and host interconnections. *J. Comp. Neurol.* 240(1): 60-70.

Martin M.J., Muotri A. *et al.* (2005). Human embryonic stem cells express an immunogenic nonhuman sialic acid. *Nat. Med.* 11(2): 228-232.

McLaren A. (2001). Ethical and social considerations of stem cell research. *Nature* 414(6859): 129-131.

Merrick J.M., Zadarlik K., Milgrom F. (1978). Characterization of the Hanganutziu-Deicher (serum-sickness) antigen as gangliosides containing n-glycolylneuraminic acid. *Int. Arch. Allergy Appl. Immunol.* 57(5): 477-480.

Meissner A., Jaenisch R. (2006). Generation of nuclear transfer-derived pluripotent ES cells from cloned Cdx2-deficient blastocysts. *Nature* 439(7073): 212-215.

Melton D.A., Daley G.Q., Jennings C.G. (2004). altered nuclear transfer in stem-cell research - a flawed proposal. *N. Engl. J. Med.* 351(27): 2791-2792.

Mitalipova M.M., Rao R.R. *et al.* (2005). Preserving the genetic integrity of human embryonic stem cells. *Nat. Biotechnol.* 23(1): 19-20.

Muchmore E.A., Diaz S., Varki A. (1998). A structural difference between the cell surfaces of humans and the great apes. *Am. J. Phys. Anthropol.* 107(2): 187-198.

Murrell W., Feron F. *et al.* (2005). Multipotent stem cells from adult olfactory mucosa. *Dev. Dyn. 233*(2):496-515.

Nagato M., Heike T. *et al.* (2005). Prospective characterization of neural stem cells by flow cytometry analysis using a combination of surface markers. *J. Neurosci. Res.* 80(4): 456-466.

Nakao N., Kakishita K. *et al.* (2001). Enhancement of the response to levodopa therapy after intrastriatal transplantation of autologous sympathetic neurons in patients with Parkinson disease. *J. Neurosurg.* 95(2): 275-284.

Nakayama T., Momoki-Soga T. *et al.* (2004). Efficient production of neural stem cells and neurons from embryonic stem cells. *Neuroreport* 15(3): 487-491.

Olanow C.W., Goetz C.G. *et al.* (2003). A double-blind controlled trial of bilateral fetal nigral transplantation in Parkinson's disease. *Ann. Neurol.* 54(3): 403-414.

Palfi S., Conde F. *et al.* (1998). Fetal striatal allografts reverse cognitive deficits in a primate model of Huntington disease. *Nat. Med.* 4(8): 963-966.

Palmer T.D., Rosman G.J. *et al.* (1991). Genetically modified skin fibroblasts persist long after transplantation but gradually inactivate introduced genes. *Proc. Natl. Acad. Sci. USA* 88(4): 1330-1334.

Perrier A.L., Tabar V. *et al.* (2004). Derivation of midbrain dopamine neurons from human embryonic stem cells. *Proc. Natl. Acad. Sci. USA* 101(34): 12543-12548.

Pizzo D.P., Paban V. *et al.* (2004). Long-term production of choline acetyltransferase in the CNS after transplantation of fibroblasts modified with a regulatable vector. *Brain Res. Mol. Brain Res.* 126(1): 1-13.

Polejaeva I.A., Chen S.H. *et al.* (2000). Cloned pigs produced by nuclear transfer from adult somatic cells. *Nature* 407(6800): 86-90.

Reynolds B.A., Weiss S. (1992). Generation of neurons and astrocytes from isolated cells of the adult mammalian central nervous system. *Science* 255(5052): 1707-1710.

Reynolds B.A., Rietze R.L. (2005). Neural stem cells and neurospheres--re-evaluating the relationship. *Nat. Methods* 2(5):333-336.

Rideout W.M. 3rd, Hochedlinger K. *et al.* (2002). Correction of a genetic defect by nuclear transplantation and combined cell and gene therapy. *Cell* 109(1): 17-27.

Rhind S.M., Taylor J.E. *et al.* (2003). Human cloning: can it be made safe? *Nat. Rev. Genet.* 4(11): 855-864.

Sawamoto K., Yamamoto A. *et al.* (2001). Direct isolation of committed neuronal progenitor cells from transgenic mice coexpressing spectrally distinct fluorescent proteins regulated by stage-specific neural promoters. *J. Neurosci. Res.* 65(3): 220-227.

Schwartz P.H., Bryant P.J. *et al.* (2003). Isolation and characterization of neural progenitor cells from post-mortem human cortex. *J. Neurosci. Res.* 74(6): 838-851.

Shiels P.G., Kind A.J. *et al.* (1999). Analysis of telomere lengths in cloned sheep. *Nature* 399(6734): 316-317.

Shihabuddin L.S., Horner P.J. *et al.* (2000). Adult spinal cord stem cells generate neurons after transplantation in the adult dentate gyrus. *J. Neurosci.* 20(23): 8727-8735.

Shin T., Kraemer D. *et al.* (2002). A cat cloned by nuclear transplantation. *Nature* 415(6874): 859.

Stenevi U., Bjorklund A., Svendgaard N.A. (1976). Transplantation of central and peripheral monoamine neurons to the adult rat brain: techniques and conditions for survival. *Brain Res.* 114(1): 1-20.

Studer L., Tabar V., McKay R.D. (1998). Transplantation of expanded mesencephalic precursors leads to recovery in parkinsonian rats. *Nat. Neurosci.* 1(4): 290-295.

Subramanian T., Emerich D.F. *et al.* (1997). Polymer-encapsulated PC-12 cells demonstrate high-affinity uptake of dopamine in vitro and 18F-Dopa uptake and metabolism after intracerebral implantation in nonhuman primates. *Cell. Transplant.* 6(5): 469-477.

Tangvoranuntakul P., Gagneux P. *et al.* (2003). Human uptake and incorporation of an immunogenic nonhuman dietary sialic acid. *Proc. Natl. Acad. Sci. USA* 100(21): 12045-12050.

Taupin P. (2005). consideration of adult neurogenesis from basic science to therapy. *Med. Sci. Monit.* 11(10): LE16-17.

Taupin P. (2006a). Derivation of embryonic stem cells for cellular therapy: Challenges and new strategies. *Med. Sci. Monit.* 12(4): RA75-78.

Taupin P. (2006b). HuCNS-SC (StemCells). *Curr. Opin. Mol. Ther.* 8(2): 156-163.

Taupin P., Gage F.H. (2002). Adult neurogenesis and neural stem cells of the central nervous system in mammals. *J. Neurosci. Res.* 69(6): 745-749.

Thomson J.A., Itskovitz-Eldor J. *et al.* (1998). Embryonic stem cell lines derived from human blastocysts. *Science.* 282(5391): 1145-1147. Erratum in: (1998) *Science* 282(5395): 1827.

Tian X.C., Xu J., Yang X. (2000). Normal telomere lengths found in cloned cattle. *Nat. Genet.* 26(3): 272-273.

Tornatore C., Baker-Cairns B. *et al.* (1996). Expression of tyrosine hydroxylase in an immortalized human fetal astrocyte cell line; in vitro characterization and engraftment into the rodent striatum. *Cell. Transplant.* 5(2): 145-163.

Tropepe V., Hitoshi S. *et al.* (2001). Direct neural fate specification from embryonic stem cells: a primitive mammalian neural stem cell stage acquired through a default mechanism. *Neuron* 30(1): 65-78.

Tuszynski M.H., Roberts J. *et al.* (1996). Gene therapy in the adult primate brain: intraparenchymal grafts of cells genetically modified to produce nerve growth factor prevent cholinergic neuronal degeneration. *Gene Ther.* 3(4): 305-314.

Tuszynski M.H., Thal L. *et al.* (2005). A phase 1 clinical trial of nerve growth factor gene therapy for Alzheimer disease. *Nat. Med.* 11(5): 551-555.

Uchida N., Buck D.W. *et al.* (2000). Direct isolation of human central nervous system stem cells. *Proc. Natl. Acad. Sci. USA* 97(26): 14720-14725.

Unsicker K., Finotto S., Krieglstein K. (1997). Generation of cell diversity in the peripheral autonomic nervous system: the sympathoadrenal cell lineage revisited. *Ann. Anat.* 179(6): 495-500.

Wakayama T., Shinkai Y. *et al.* (2000). Cloning of mice to six generations. *Nature* 407(6802): 318-319.

Wakitani S., Takaoka K. *et al.* (2003). Embryonic stem cells injected into the mouse knee joint form teratomas and subsequently destroy the joint. *Rheumatology (Oxford).* 42(1): 162-165.

Wei M.Q., Lejnieks D.V. *et al.* (1999). Sustained gene expression in transplanted skin fibroblasts in rats. *Gene Ther.* 6(5): 840-844.

Weiss R.A. (1998). Xenotransplantation. *BMJ* 317(7163): 931-934.

Wilmut I. (2003). Dolly-her life and legacy. *Cloning Stem Cells* 5(2): 99-100.

Winkler J., Suhr S.T. *et al.* (1995). Essential role of neocortical acetylcholine in spatial memory. *Nature* 375(6531): 484-487.

Wobus A.M., Boheler K.R. (2005). Embryonic stem cells: prospects for developmental biology and cell therapy. *Physiol. Rev.* 85(2): 635-678.

Xu R.H., Peck R.M. *et al.* (2005). Basic FGF and suppression of BMP signaling sustain undifferentiated proliferation of human ES cells. *Nat. Methods* 2(3): 185-190.

Zhang S.C., Wernig M. *et al.* (2001). In vitro differentiation of transplantable neural precursors from human embryonic stem cells. *Nat. Biotechnol.* 19(12): 1129-1133.

Chapter VIII

Stem Cells Engineering for Cell-Based Therapy

Abstract

Stem cells carry the promise to cure a broad range of diseases and injuries, from diabetes, heart and muscular diseases, to neurological diseases, disorders and injuries. Significant progresses have been made in stem cell research over the past decade; the derivation of embryonic stem cells (ESCs) from human tissues, the development of cloning technology by somatic cell nuclear transfer (SCNT) and the confirmation that neurogenesis occurs in the adult mammalian brain and that neural stem cells (NSCs) reside in the adult central nervous system (CNS), including that of humans. Despite these advances, there may be decades before stem cell research will translate into therapy. Stem cell research is also subject to ethical and political debates, controversies and legislation, which slow its progress. Cell engineering has proven successful in bringing genetic research to therapy. In this review, I will review, in two examples, how investigators are applying cell engineering to stem cell biology to circumvent stem cells' ethical and political constraints and bolster stem cell research and therapy.

Introduction

Cellular therapy is the replacement of lost or dysfunctional tissues by new ones. Various cell types have been evaluated and considered for therapy. In the CNS, fetal neuronal tissue has been particularly evaluated for its merit to treat neurological diseases and injuries [1]. While numerous experimental and clinical transplantation studies showed that fetal neuronal transplants improve functional deficits in models of CNS diseases [2–5], others reported less positive outcomes [6, 7]. In addition, the rate of survival of fetal neuronal cells transplanted in the adult brain is relatively low, requiring large quantities of tissue, generally from several fetuses, for therapy. Researchers are looking at other opportunities for cellular therapy, particularly in the CNS.

Stem cells are undifferentiated cells that have the ability to self-renew and differentiate into other cell types. Stem cells represent a promising model for cell-based therapy [8]. ESCs are self-renewing pluripotent cells derived from the inner cell mass (ICM) of the blastocyst. ESCs can generate cells derived from the three primary germ layers; ectoderm, mesoderm and endoderm. Because ESCs can give rise to the various cell types of the body, an estimated 220 cell types in humans, they carry the hope of curing a broad range of diseases and injuries [9]. Stem cells exist in fetal and adult tissues. These stem cells are multipotent; they generate cells from the tissue from which they are derived [8]. In the CNS, with the confirmation that neurogenesis occurs in the adult brain and NSCs reside in the adult CNS, the adult brain appears to have the ability of limited self-repair [10–12]. Neural stem and progenitor cells have been isolated from adult tissues [13, 14], including from biopsies and *post-mortem* tissues [15, 16], providing a potential source of tissues for the treatment of neurological diseases and injuries. Neural stem and progenitor cells derived from human fetal tissues are currently in clinical trial for the treatment of Batten's disease, a childhood neurodegenerative disorder [17]. Preclinical data show that grafted neural stem and progenitor cells survive in damaged brain tissues, migrate to specific sites of degeneration where they differentiate into neural lineages, and have a beneficial effect on functional recovery [17]. This emphasizes the potential of neural stem and progenitor cells for therapy.

There are ethical considerations for the use of stem cells for research and therapy. ESCs have been derived from the ICM of human embryos (hECSs) [18]. hESCs are derived from leftover frozen embryos, created for *in vitro* fertilization, destined to be discarded. Because it involves the destruction of embryos, there is a debate over the use of hESCs for research and therapy. On the one hand, there are scientific reports and evidence that ESCs, including hESCs, differentiate into various cell types of the body, such as neuronal and insulin producing cells [19], backing the scientific claim for the potential of ESCs for therapy. On the other hand, opponents of the use of hESCs for research and therapy argue that it is morally not acceptable to destroy human life. The scientific and ethical debate over the use of hESCs for research and therapy considerably slows stem cell research.

It is proposed to use fetal or adult tissue-derived stem cells to circumvent the ethical issues over the use of ESCs for research and therapy. However, because fetal and adult stem cells are multipotent, they would have a more limited potential for the treatment of diseases and injuries. Recent evidence suggests that adult tissue-derived stem cells may have a broader potential; they may have the capacity to give rise to cells from other lineages than the one from which they are derived [20]. This would give adult stem cells a broader spectrum to treat diseases and injuries [21]. However, data related to the broader potential of adult tissue-derived stem cells are the source of debates and controversies, and remain to be validated therapeutically. More recently, stem cells extracted from the amniotic fluid were reported to have similar potential to ESCs, fueling the search for an alternative source of cells for therapy without the controversies over the ESCs for clinical research and therapy [22]. Although multipotent stem cells offer an alternative to ESCs for therapy, there are also ethical considerations to take into account when using fetal and adult human-derived tissue for therapy. Among them validating the use of human tissue-derived cells, including fetal and adult stem cells, involves experimentation where human-derived cells are transplanted into animals— with different ratios of human versus host cells—generating chimeras [23]; experiments that are subject to strict regulations and controversies [24]. Other issues, like possible non-ethical origin of tissue (e.g., absence of consent from the donor), can potentially be damaging for research involving human tissues, and particularly for stem cell research [25]. There is also serious ethical consideration regarding fetal sources of stem cells (versus hESCs). While the use of hESCs is morally not acceptable because it involves the destruction of embryos, the use of fetuses is morally acceptable when it involves tissues originating from spontaneous abortions (miscarriages) with the consent from the parents, as opposed to elective abortions. In all, strict ethical and political guidelines must be followed when using human tissues for therapy and particularly for stem cell research.

Transgenic modifications offer the ability to modify the cells' genome to accomplish specific functions. It has been instrumental in the study of gene functions, and as a therapeutic tool to produce biologically active substances, such as neurotransmitter synthesizing enzymes and trophic factors [26, 27]. Stem cells have properties that make them a suitable candidate for benefiting from gene therapy. Though stem cells can give rise to the various cells types of the body, protocols to differentiate them in a wide variety and high yield of cell types, including neuronal and insulin producing cells, remain to be established. In the CNS, beside the replacement of lost or degenerated nerve cells by the grafted cells, studies revealed that grafted neural stem and progenitor cells may also promote functional recovery through the secretion of trophic factors [28–30]. This shows that transgenic expression of trophic factors or expression of missing proteins is increasingly likely to play a role in cell-based therapies. Stem cells, and particularly NSCs, have inherent properties to migrate to tumors, injured and diseased sites after transplantation [31, 32]. The properties of NSCs to be genetically modified [14] and to migrate to diseased or degenerated sites provide unique opportunities to target these areas and provide missing proteins to promote recovery. Genetically modifying stem cells has also been proposed to circumvent some of the ethical issues associated with the use of ESCs for clinical research and therapy. In this review, I will discuss recent studies involving stem cells engineered to bolster stem cell research and therapy; their potentials, limitations and controversies.

Cell Engineering to Derive ESCs without the Destruction of Embryos

SCNT is a cloning strategy, originally reported by Campbell et al. [33], in which nuclei are isolated from a donor's somatic cells, such as fibroblasts, and are transferred into enucleated oocytes from female donors [33]. By mechanisms yet to be uncovered, the cytoplasm of the oocytes reprograms the chromosomes of the somatic cell nucleus and the cloned cells develop into blastocysts, from which ESCs can be derived [34]. One of the landmarks of SCNT is the potential to generate isogeneic ESCs, carrying a set of chromosomes identical to that of an individual, and therefore unlikely to be rejected after transplantation into that individual [35].

Beside its potential, human cloning is the source of scientific, ethical and political debates, controversies and legislation [36–38]. SCNT has been applied to clone various animals, to derive ESCs from various species, but remains to be applied successfully to derive hESCs [39]. The General Assembly of the United Nations has adopted a declaration calling on governments to ban all forms of human cloning that are 'incompatible with human dignity and the protection of human life'. In the US, human cellular cloning (that is therapeutic cloning, using SCNT) is not banned, while human being cloning (i.e., reproductive cloning) is banned. Countries like England, and more recently Australia, with the aim to promote stem cell research, allow therapeutic cloning, but not reproductive cloning.

In an attempt to circumvent the ethical and political debate over the use SCNT for clinical research, Meissner and Jaenisch [40] used a variation of SCNT, altered nuclear transfer (ANT), to derive ESCs in mice [40]. ANT is a variation of SCNT proposed by W B Hurlbut [41]. In ANT, a gene crucial for trophectoderm development, the gene CDX2, is inactivated *in vitro*, in the donor cells. CDX2 encodes the earliest-known trophectoderm-specific transcription factor that is activated in the eight-cell embryos, and is essential for establishment and function of the trophectoderm. Inactivating the gene CDX2 eliminates formation of the fetal–maternal interface, but spares the ICM from which ESCs could be derived. The nucleus deficient in CDX2 is then transferred into enucleated oocytes from female donors, and submitted to the same protocols as for SCNT. Because the eggs created from nuclei deficient in CDX2 produce embryos that are unable to implant into the uterus, and do not pursue their developments, ANT has been proposed as a variation of nuclear transfer to derive ESCs, without the destruction of embryos. Meissner and Jaenisch genetically modified the donor cells, mouse fibroblasts, by inserting in their genome a cassette coding for RNAi cdx2 and the green fluorescent protein (GFP), flanked by two LoxP sequences [40]. The cassette was inserted in the cells' genome using a lenti viral vector. The nuclei of genetically engineered fibroblasts, selected by means of GFP fluorescence, were transferred into enucleated oocytes, and submitted to the same protocols as for SCNT. The eggs divided, produced cloned blastocysts that were morphologically abnormal and lacked functional trophoblasts. The cloned blastocysts did not implant into the uterus, and ESCs could be derived from their ICMs. To maintain the developmental potential of the generated ESCs, the expression of CDX2 is re-established by deleting the cassette RNAi cdx2, using a plasmid coding for Cre-recombinase. This study was the first to report the derivation of ESCs by ANT.

There is a debate as to whether a mutant embryo is equivalent to a normal embryo. Proponents of ANT claim that because the eggs created from nuclei deficient in CDX2 produce embryos that are unable to implant into the uterus and do not undergo subsequent development, ANT represents a variation of nuclear transfer to derive ESCs without the destruction of embryos [42]. Alternatively, it has been argued that finding it acceptable to destroy a CDX2 mutant embryo, but not a normal embryo, is a 'flawed proposal', as there is no basis for concluding that the action of CDX2, or any other gene, represents a transition point at which a human embryo acquires moral status [43]. So whether ANT solves the issues of the derivation of ESCs without destruction of embryos remains a source of debate [44–47]. Further, in the study by Meissner and Jaenisch [40], it is not known whether cloned ESCs with an inactivated gene CDX2 have the same developmental potential as ESCs derived from donated eggs. Studies have reported that SCNT may affect the developmental potential of cloned animals and ESCs [48–50]. Meissner and Jaenisch [40] also used a lenti virus to engineer the donor cells [40]. All of which may affect the developmental and therapeutic potential of ESCs generated by ANT. Nonetheless, this study highlights the potential of cell engineering for the advancement of research in stem cell biology.

Adult Neural Stem Cell Engineering

Contrary to a long-held belief, neurogenesis occurs in the adult mammalian brain, including in humans [10, 11]. Neurogenesis occurs primarily in two areas of the adult brain, the dentate gyrus of the hippocampus and the subventricular zone along the ventricles. It is hypothesized that newly generated neuronal cells originate from stem cells in the adult brain [12]. NSCs are the self-renewing multipotent cells that generate the main cell types of the nervous system, neurons, astrocytes and oligodendrocytes. Neural stem and progenitor cells have been isolated and characterized *in vitro* from various regions of the adult CNS, including the spinal cord, supporting the existence of NSCs in the CNS [51]. The generation of new neuronal cells in the adult brain and the isolation and characterization of neural stem and progenitor cells from the adult CNS suggest that the adult brain may be amenable to repair. Cell therapy in the adult CNS could involve the stimulation of endogenous neural stem or progenitor cells or the transplantation of adult-derived neural stem and progenitor cells. Adult-derived neural stem and progenitor cells have been transplanted in animal models, and shown functional engraftment [14, 52, 53]. More recently, grafted human-derived neural stem and progenitor cells show functional integration and promote functional recovery in an animal model of spinal cord injury, supporting the potential of neural stem and progenitor cells for therapy [54].

Adult neural stem and progenitor cells can be genetically modified by retroviral-mediated transfection, rendering them a vehicle for gene therapy [14, 52]. Adult-derived neural stem and progenitor cells genetically modified to express acid sphingomyelinase can lead to a reversal of lysosomal storage pathology when transplanted into animal models of Niemann-Pick's disease [55]. Niemann-Pick's disease is a lysosomal storage disorder in which deficiency of acid sphingomyelinase leads to the intracellular accumulation of sphingomyelin and cholesterol in lysosomes. This highlights the potential of genetically modified NSCs for the treatment of lysosomal storage diseases and other genetic diseases of the CNS, but also for delivering trophic factors for the treatment of neurodegenerative diseases.

The relevance of genetically modified NSCs for stem cell therapy is further highlighted by the potential of NSCs for the treatment of brain tumors. Neural stem and progenitor cells migrate to tumors and injured and diseased sites when transplanted in the CNS, either by systemic injection, or through the cerebrospinal fluid [31, 32, 56–59]. The injected cells migrate to the diseased or degenerated areas where they integrate with the host tissue. The properties of NSCs to be genetically engineered and to migrate to tumor sites have been proposed for the treatment of brain tumors. It is proposed to genetically engineer NSCs with 'suicide genes', such as genes coding for cytolytic activities or anti-tumor cytokines, to attack and destroy brain tumor cells [60, 61]. This further extends the potential of genetically modified stem cells for cancer therapy, particularly in the CNS.

Conclusion

Stem cell therapy holds the promise to treat a broad range of diseases and injuries. The promise of stem cell therapy, particularly in the CNS, is to regenerate and reconstruct the original pathway to promote functional recovery, but it may be years away before it emerges as a viable therapy. Genetically engineering cells has proven valuable to understand gene function, and to deliver missing trophic factors or neurotransmitter synthesizing enzymes in the CNS. The studies reviewed show that genetically engineering stem cells, and particularly NSCs, may offer an opportunity to bolster stem cell research and therapy. There are however several limitations for the application of gene therapy-based strategies for therapy. Among them, the long-term expression of the transgenes, and the risk and limitations of using genetically engineered cells for therapy.

Acknowledgments

Reproduced with the permission of IOP Publishing Ltd.: Taupin P. Stem cells engineering for cell-based therapy. Journal of Neural Engineering (2007) 4(3): R59-63.

References

[1] Bjorklund A, Dunnett SB, Brundin P, Stoessl AJ, Freed CR, Breeze RE, Levivier M, Peschanski M, Studer L, Barker R. (2003) Neural transplantation for the treatment of Parkinson's disease. *Lancet Neurol.* 2, 437-45.

[2] Gage FH, Bjorklund A, Stenevi U, Dunnett SB, Kelly PA. (1984) Intrahippocampal septal grafts ameliorate learning impairments in aged rats. *Science* 225, 533-6.

[3] Palfi S, Conde F, Riche D, Brouillet E, Dautry C, Mittoux V, Chibois A, Peschanski M, Hantraye P. (1998) Fetal striatal allografts reverse cognitive deficits in a primate model of Huntington disease. *Nat. Med.* 4, 963-6.

[4] Studer L, Tabar V, McKay RD. (1998) Transplantation of expanded mesencephalic precursors leads to recovery in parkinsonian rats. *Nat. Neurosci.* 1, 290-5.

[5] Freed C R, Greene P E, Breeze R E, Tsai W Y, DuMouchel W, Kao R, Dillon S, Winfield H, Culver S, Trojanowski J Q, Eidelberg D, Fahn S. (2001) Transplantation of embryonic dopamine neurons for severe Parkinson's disease. *N. Engl. J. Med.* 344, 710-9.

[6] Olanow CW, Goetz CG, Kordower JH, Stoessl AJ, Sossi V, Brin MF, Shannon KM, Nauert GM, Perl DP, Godbold J, Freeman TB. (2003) A double-blind controlled trial of bilateral fetal nigral transplantation in Parkinson's disease. *Ann. Neurol.* 54, 403-14.

[7] Bachoud-Levi AC, Gaura V, Brugieres P, Lefaucheur JP, Boisse MF, Maison P, Baudic S, Ribeiro MJ, Bourdet C, Remy P, Cesaro P, Hantraye P, Peschanski M. (2006) Effect of fetal neural transplants in patients with Huntington's disease 6 years after surgery: a long-term follow-up study *Lancet Neurol.* 5, 303-9.

[8] Potten CS, Loeffler M. (1990) Stem cells: attributes, cycles, spirals, pitfalls and uncertainties. Lessons for and from the crypt. *Development* 110, 1001-20.

[9] Wobus AM, Boheler KR. (2005) Embryonic stem cells: prospects for developmental biology and cell therapy. *Physiol. Rev.* 85, 635-78.

[10] Gage FH. (2000) Mammalian neural stem cells. *Science* 287, 1433-8.

[11] Kaplan MS. (2000) Environment complexity stimulates visual cortex neurogenesis: death of a dogma and a research career. *Trends Neurosci.* 24, 617-20.

[12] Taupin P, Gage FH. (2002) Adult neurogenesis and neural stem cells of the central nervous system in mammals. *J. Neurosci. Res.* 69, 745-9.

[13] Reynolds BA, Weiss S. (1992) Generation of neurons and astrocytes from isolated cells of the adult mammalian central nervous system. *Science* 255, 1707-10.

[14] Gage FH, Coates PW, Palmer TD, Kuhn HG, Fisher LJ, Suhonen JO, Peterson DA, Suhr ST, Ray J. (1995) Survival and differentiation of adult neuronal progenitor cells transplanted to the adult brain *Proc. Natl. Acad. Sci. USA* 92, 11879-83.

[15] Roy NS, Wang S, Jiang L, Kang J, Benraiss A, Harrison-Restelli C, Fraser RA, Couldwell WT, Kawaguchi A, Okano H, Nedergaard M, Goldman SA.(2000) In vitro neurogenesis by progenitor cells isolated from the adult human hippocampus. Nat. Med. 6, 271-7.

[16] Palmer TD, Schwartz PH, Taupin P, Kaspar B, Stein SA, Gage FH. (2001) Cell culture. Progenitor cells from human brain after death. Nature 411, 42-3.

[17] Taupin P. (2006) HuCNS-SC (StemCells). Curr. Opin. Mol. Ther. 8, 156-63.

[18] Thomson JA, Itskovitz-Eldor J, Shapiro SS, Waknitz MA, Swiergiel JJ, Marshall VS, Jones JM. (1998) Embryonic stem cell lines derived from human blastocysts. *Science.* 282, 1145-7. Erratum in: (1998) *Science* 282, 1827.

[19] Reubinoff BE, Pera MF, Fong CY, Trounson A, Bongso A. (2000) Embryonic stem cell lines from human blastocysts: somatic differentiation in vitro. *Nat. Biotechnol.* 18, 399-404. Erratum in: (2000) *Nat. Biotechnol.* 18, 559.

[20] D'Amour KA, Gage FH. (2002) Are somatic stem cells pluripotent or lineage-restricted? *Nat. Med.* 8, 213-4.

[21] Taupin P. (2006) Adult neural stem cells, neurogenic niches and cellular therapy. *Stem Cell Reviews* 2, 213-9.

[22] De Coppi P, Bartsch G Jr, Siddiqui MM, Xu T, Santos CC, Perin L, Mostoslavsky G, Serre AC, Snyder EY, Yoo JJ, Furth ME, Soker S, Atala A. (2007) Isolation of amniotic stem cell lines with potential for therapy. *Nat. Biotechnol.* 25, 100-6.

[23] Muotri AR, Nakashima K, Toni N, Sandler VM, Gage FH. (2005) Development of functional human embryonic stem cell-derived neurons in mouse brain. *Proc. Natl. Acad. Sci. USA* 102, 18644-8.

[24] Karpowicz P, Cohen CB, van der Kooy D. (2004) It is ethical to transplant human stem cells into nonhuman embryos. *Nat. Med.* 10, 331-5.

[25] Holden C. (2005) Stem cell research. Korean cloner admits lying about oocyte donations. *Science* 310, 1402-3.

[26] Gage FH, Kawaja MD, Fisher LJ. (1991) Genetically modified cells: applications for intracerebral grafting. *Trends Neurosci.* 14, 328-33.

[27] Mulligan RC. (1993) The basic science of gene therapy. *Science* 260, 926-32.

[28] Lu P, Jones LL, Snyder EY, Tuszynski MH. (2003) Neural stem cells constitutively secrete neurotrophic factors and promote extensive host axonal growth after spinal cord injury. *Exp. Neurol.* 181, 115-29.

[29] Llado J, Haenggeli C, Maragakis NJ, Snyder EY, Rothstein JD. (2004) Neural stem cells protect against glutamate-induced excitotoxicity and promote survival of injured motor neurons through the secretion of neurotrophic factors. *Mol. Cell Neurosci.* 27, 322-31.

[30] Bjugstad KB, Redmond DE Jr, Teng YD, Elsworth JD, Roth RH, Blanchard BC, Snyder EY, Sladek JR Jr. (2005) Neural stem cells implanted into MPTP-treated monkeys increase the size of endogenous tyrosine hydroxylase-positive cells found in the striatum: a return to control measures *Cell Transplant.* 14, 183-92.

[31] Zhang RL, Zhang L, Zhang ZG, Morris D, Jiang Q, Wang L, Zhang LJ, Chopp M. (2003) Migration and differentiation of adult rat subventricular zone progenitor cells transplanted into the adult rat striatum. *Neuroscience* 116, 373-82.

[32] Burnstein RM, Foltynie T, He X, Menon DK, Svendsen CN, Caldwell MA. (2004) Differentiation and migration of long term expanded human neural progenitors in a partial lesion model of Parkinson's disease. *Int. J. Biochem. Cell Biol.* 36, 702-13.

[33] Campbell KH, McWhir J, Ritchie WA, Wilmut I. (1996) Sheep cloned by nuclear transfer from a cultured cell line. *Nature* 380, 64-6.

[34] Wakayama T. (2006) Establishment of nuclear transfer embryonic stem cell lines from adult somatic cells by nuclear transfer and its application. Ernst Schering Res Found Workshop. 60, 111-23.

[35] Lanza RP, Chung HY, Yoo JJ, Wettstein PJ, Blackwell C, Borson N, Hofmeister E, Schuch G, Soker S, Moraes CT, West MD, Atala A. (2002) Generation of histocompatible tissues using nuclear transplantation. *Nat. Biotechnol.* 20, 689-96.

[36] Trounson A, Pera M. (19980. Potential benefits of cell cloning for human medicine. *Reprod. Fertil. Dev.* 10, 121-5.

[37] Lanza RP, Cibelli JB, West MD. (1999) Human therapeutic cloning. *Nat. Med.* 5, 975-7.

[38] Jaenisch R, Wilmut I. (2001) Developmental biology. Don't clone humans! *Science* 291, 2552.

[39] Kennedy D. (2006) Editorial retraction. *Science* 311, 335.

[40] Meissner A, Jaenisch R. (2006) Generation of nuclear transfer-derived pluripotent ES cells from cloned Cdx2-deficient blastocysts. *Natur.* 439, 212-5.

[41] Hurlbut WB. (2005) Altered nuclear transfer. *N. Engl. J. Med.* 352, 1153-4.

[42] Hurlbut WB. (2005) Altered nuclear transfer as a morally acceptable means for the procurement of human embryonic stem cells. *Perspect Biol. Med.* 48, 211-28.

[43] Melton DA, Daley GQ, Jennings CG. (2004) Altered nuclear transfer in stem-cell research - a flawed proposal. *N. Engl. J. Med.* 351, 2791-2.

[44] Weissman IL. (2006) Medicine: politic stem cells. *Nature* 439, 145-7.

[45] Solter D. (2005) Politically correct human embryonic stem cells? *N. Engl. J. Med.* 353, 2321-3.

[46] Jaenisch R, Meissner A. (2006) Politically correct human embryonic stem cells? *N. Engl. J. Med.* 354, 1208-9.

[47] Taupin P. (2006) Derivation of embryonic stem cells. *Med. Sci.* (Paris). 22, 478-80.

[48] Shiels PG, Kind AJ, Campbell KH, Waddington D, Wilmut I, Colman A, Schnieke AE. (1999) Analysis of telomere lengths in cloned sheep. *Nature* 399, 316-7.

[49] Rhind SM, Taylor JE, De Sousa PA, King TJ, McGarry M, Wilmut I. (2003) Human cloning: can it be made safe? *Nat. Rev. Genet.* 4, 855-64.

[50] Wakayama S, Jakt ML, Suzuki M, Araki R, Hikichi T, Kishigami S, Ohta H, Van Thuan N, Mizutani E, Sakaide Y, Senda S, Tanaka S, Okada M, Miyake M, Abe M, Nishikawa S, Shiota K, Wakayama T. (2006) Equivalency of nuclear transfer-derived embryonic stem cells to those derived from fertilized mouse blastocysts. *Stem Cells* 24, 2023-33.

[51] Taupin P. (2006) Neural Progenitor and Stem Cells in the Adult Central Nervous System. Annals Academy of Medicine Singapore. 35, 814-7.

[52] Shihabuddin LS, Horner PJ, Ray J, Gage FH. (2000) Adult spinal cord stem cells generate neurons after transplantation in the adult dentate gyrus. *J. Neurosci.* 20, 8727-35.

[53] Suhonen JO, Peterson DA, Ray J, Gage FH. (1996) Differentiation of adult hippocampus-derived progenitors into olfactory neurons in vivo. *Nature* 383, 624-7.

[54] Cummings BJ, Uchida N, Tamaki SJ, Salazar DL, Hooshmand M, Summers R, Gage FH, Anderson AJ. (2005) Human neural stem cells differentiate and promote locomotor recovery in spinal cord-injured mice. *Proc Natl Acad Sci. USA* 102, 14069-74.

[55] Shihabuddin LS, Numan S, Huff MR, Dodge JC, Clarke J, Macauley SL, Yang W, Taksir TV, Parsons G, Passini MA, Gage FH, Stewart GR. (2004) Intracerebral transplantation of adult mouse neural progenitor cells into the Niemann-Pick-A mouse leads to a marked decrease in lysosomal storage pathology. *J. Neurosci.* 24, 10642-51.

[56] Aboody KS, Brown A, Rainov NG, Bower KA, Liu S, Yang W, Small JE, Herrlinger U, Ourednik V, Black PM, Breakefield XO, Snyder EY. (2000) Neural stem cells display extensive tropism for pathology in adult brain: evidence from intracranial gliomas. Proc Natl Acad Sci. USA 97, 12846-51. Erratum in: (2001) *Proc. Natl. Acad. Sci. USA* 98, 777.

[57] Veizovic T, Beech JS, Stroemer RP, Watson WP, Hodges H. (2001) Resolution of stroke deficits following contralateral grafts of conditionally immortal neuroepithelial stem cells. *Stroke* 32, 1012-9.

[58] Fujiwara Y, Tanaka N, Ishida O, Fujimoto Y, Murakami T, Kajihara H, Yasunaga Y, Ochi M. (2004) Intravenously injected neural progenitor cells of transgenic rats can migrate to the injured spinal cord and differentiate into neurons, astrocytes and oligodendrocytes. *Neurosci. Lett.* 366, 287-91.

[59] Pluchino S, Quattrini A, Brambilla E, Gritti A, Salani G, Dina G, Galli R, Del Carro U, Amadio S, Bergami A, Furlan R, Comi G, Vescovi AL, Martino G. (2003) Injection of adult neurospheres induces recovery in a chronic model of multiple sclerosis. *Nature* 422, 688-94.

[60] Yip S, Aboody KS, Burns M, Imitola J, Boockvar JA, Allport J, Park KI, Teng YD, Lachyankar M, McIntosh T, O'Rourke DM, Khoury S, Weissleder R, Black PM, Weiss W, Snyder EY. (2003) Neural stem cell biology may be well suited for improving brain tumor therapies *Cancer J.* 9, 189-204.

[61] Shah K, Bureau E, Kim DE, Yang K, Tang Y, Weissleder R, Breakefield XO. (2005) Glioma therapy and real-time imaging of neural precursor cell migration and tumor regression. *Ann. Neurol.* 57, 34-41.

Chapter IX

Therapeutic Neuronal Stem Cells: Patents at the Forefront

Abstract

Background: Neural stem cells (NSCs) hold the promise to cure a broad range of neurological diseases and injuries, particularly neurodegenerative diseases and spinal cord injuries. The recent confirmation that neurogenesis occurs in the adult brain and that NSCs reside in the adult CNS opens new opportunities and avenues for cellular therapy. Objectives: To provide an overview of the current patent situation related to NSCs and to highlight the limitations and challenges of bringing NSC research to therapy. Method: Reviewing the early studies and patents in NSC research. Conclusion: NSCs are in clinical trials for Batten and Parkinson's diseases. However, clinical development and other limitations will make it difficult for pharmaceutical/biotech companies to break even with these early patents.

1. Background

Neural stem cells (NSC) are the self-renewing multipotent cells that generate the main phenotypes of the nervous system, neurons, astrocytes and oligodendrocytes [1]. They hold the potential to treat and cure a broad range of neurological diseases and injuries, ranging from neurodegenerative diseases, like Alzheimer and Parkinson's diseases, to retinal diseases, cerebral strokes and spinal-cord injuries. Neurogenesis in the adult mammalian brain occurs primarily in two regions, the dentate gyrus (DG) of the hippocampus and subventricular zone (SVZ), in various species including human [2,3]. It is hypothesized that newborn neuronal cells in the adult brain originate from residual stem cells. In support of this contention, neural progenitor and stem cells have been isolated and characterized in vitro, from various regions of the adult CNS [4]. The confirmation that neurogenesis occurs in the adult mammalian brain and that NSCs reside in the adult mammalian CNS has tremendous consequences for our understanding of developmental biology and therapy. The adult brain has the potential to, and may be amenable to, repair. Therapeutic interventions may involve the stimulation of endogenous neural progenitor or stem cells, and the transplantation of adult-derived neural progenitor and stem cells.

2. Neural Stem Cells, Therapeutic Potential and Patents

The advent of adult neurogenesis and NSC research has been associated with the filing and granting of patents applications. The criteria for selection of the patents cited was to identify some of the early patents of the field of NSC research, on which several pharmaceutical and biotech companies have been building and which have resulted in clinical trials. Patent WO/1993/001275, assigned to S Weiss and BA Reynolds, and first disclosed in 1993, lays the ground for protection of the intellectual property, for the preparation and use of neural progenitor and stem cells *in vitro*, *ex vivo* and *in vivo* [5]. With other patents, including patents WO/1995/013364, WO/1996/015224 and WO/1996/015226, it claims the use of neural progenitor and stem cells, and their progeny, isolated as neurospheres of various ages and from various species, from either normal or diseased CNS tissue, as a model to study neural development, function, to screen the effect of biological agents and to develop novel therapies [6-8]. These patents are based on the published work by Reynolds and Weiss [9], reporting the isolation and characterization of the neural progenitor and stem cells from the fetal and adult mammalian tissues [4,10]. In 2001, Palmer and collaborators reported the isolation and characterization of neural progenitor cells from human postmortem tissues [11]. The inventors filed for a patent application, patent WO/2002/036749, to claim their work; the culture and expansion of stem cells from postmortem tissues for therapy, and methodologies to isolate and propagate stem cells from biopsies and postmortem tissues [12].

Patents have been filed for methods to improve culture conditions, like promoting survival, proliferation and differentiation of neural progenitor and stem cells in vitro. Patent WO/2000/050572 covers a method for the *in vitro* proliferation of NSC cultures, using a growth factor and collagenase [13]. Collagenase is used as a cell-dissociating agent; its use improves cell isolation, viability and proliferation. Patent WO/2000/033791 relates to the isolation, purification and use of the co-factor of fibroblast growth factor-2 (FGF-2), glycosylated cystatin C (CCg) [14]. FGF-2 requires CCg for its mitogenic activity on self-renewing multipotent NSCs in vitro, from single cells, to stimulate neurogenesis in vivo, and to promote the growth and expansion of neural progenitor and stem cells, isolated and cultured for human postmortem tissues [11,15]. These factors and conditions may contribute to promote the use of NSCs for therapy.

Fetal and adult stem cells are multipotents; they generate lineage-specific cell types restricted to the tissues from which they are derived. Embryonic stem cells are pluripotents; they generate cells of the three germ layers – ectoderm, mesoderm and endoderm. Neural progenitor and stem cells, isolated from the adult brain and cultured in vitro, have been reported to give rise to lineages other than neuronal, in vitro and ex vivo, particularly blood cells [16]. Stem cells isolated from adult tissues other than the brain, like the skin, blood and bone marrow, have been reported to give rise to lineages other than the one from which they are derived, particularly neuronal lineages [17-19]. It is proposed that adult stem cells would have a broader potential than previously thought [20]. This broader potential would have tremendous possibilities for cellular therapy, particularly in the CNS, as neuronal phenotypes could be generated from tissues other from than the nervous system. Of particular interest would be the generation of nerve cells from skin tissues, potentially allowing autologous transplantation and limiting potential damage to the CNS associated with the isolation of cells. Patent applications for the broader potential of adult stem cells, particularly adult NSCs to generate haematopoietic cells and non-haematopoietic lineages generated from haematopoietic stem cells, have been filed WO/1999/016863 and WO/2001/071016 [21,22]. However, the broader potential of stem cells is the source of debates and controversies, as it could originate from cell fusion, transdifferentiation or transformation, all of which could have adverse effects on their therapeutic potentials. Alternatively, lineage non-specific adult stem cells could be generated by de-differentiation or reprogramming, or by the introduction of developmental genes to generate stem or progenitor cells of a tailored phenotype [23,24].

Adult-derived neural progenitor and stem cells can be genetically modified [25] to express tropic factors or neurotransmitter-synthesizing enzymes. Patent applications on methods to genetically engineered neural progenitor and stem cells in vitro have also been submitted. These include various processes for genetically modifying neural progenitor and stem cells, including with reporter genes like the *Escherichia coli* β-galactosidase or green fluorescent protein, and the isolation of neural progenitor and stem cells from transgenic animals. Patent US-05,750,376 covers the *in vitro* growth and proliferation of genetically modified NSCs [26]. Adult neural progenitor and stem cells have been genetically engineered to express acid sphingomyelinase reverse lysosomal storage pathology in animal models of Niemann-Pick's disease [27], a strategy subsequently filed under patent WO/2006/074387 [28]. Hence, NSCs offer a tremendous potential not only for the treatment of neurodegenerative diseases and injuries but also for the treatment of genetic diseases of the CNS.

3. Expert Opinion

The advent of adult neurogenesis and NSC research offer tremendous opportunities for cellular therapy in the CNS. Neural progenitor and stem cells can be isolated either from patients themselves - allowing autologous transplantation - or from postmortem tissues, providing unlimited sources of material for transplantation. Neural progenitor and stem cells are currently in, or are being considered for, clinical trials for the treatment of Batten and Parkinson's diseases [29]. Batten disease (BD) is a rare and fatal genetic disorder that begins in childhood. In BD, various genes responsible for the breakdown of lipofuscins in the nerve cells are missing or faulty. It is proposed that grafted stem cells would migrate to the areas of neurodegeneration and compensate for the missing enzymes. Fetal-derived neural progenitor and stem cells are being utilized in the clinical trials for BD, whereas autologous transplantation is considered for the clinical trials for Parkinson's disease [30,31].

Although stem cell therapy is promising, there are limitations, pitfalls and risks to consider and monitor, particularly with the use of neural progenitor and stem cells. Among these is the fact that NSCs are still elusive; they have yet to be unequivocally identified and characterized [32,33]. Current protocols devised to isolate and culture self-renewing multipotent NSCs in vitro yield heterogeneous populations of neural progenitor and stem cells [34]. Some authors have used strategies - for example the combination of the use of antibodies against membranous molecular markers and fluorescence-activated cell sorting - specifically to isolate and purify subpopulations of cells characterized as self-renewing multipotent NSCs *in vitro*. Patents WO/2000/047762, WO/2004/020597 and US-7,381,561 disclose methods for identifying, isolating and enriching neural stem/progenitor cells, and for using antibodies to membranous markers, like CD49, CD133 CD15 as well as proprietary monoclonal antibodies [35-37]. However, these populations of cells expand as heterogeneous populations in vitro, limiting their therapeutic potential. To circumvent this, it is proposed to predifferentiate in vitro neural progenitor and stem cells toward the neuronal, astroglial or oligodendroglial lineages prior to transplantation. One of the limitations of such a strategy would concern in particular predifferentiated neuronal cells that might not integrate with the neuronal network upon transplantation. Among other limitations, pitfalls and risks to be considered are the controversies and debates over the broader potential of adult stem cells and the risks of tumour formation and inflammation that are associated with the transplantation of neural progenitor and stem cells, particularly if these neural progenitor and stem cells originate from donors whose compatibility may not be matched to those of the patient. There are also risks with autologous transplantation, as the isolation of brain tissues represents an invasive surgery, with risks to the patients [4].

The scientific and technical challenges that need to be overcome also have significant consequences for the pharmaceutical/biotech companies engaged in this field of research and development. Due to the lengthy and multiple processes involved in bringing NSC research to therapy, the 20-year protection of patents may not be sufficient for a pharmaceutical/biotech company to bring a product to therapy within a time-frame that allows it to benefit from the protection of its intellectual property. This is a particular concern for patented discoveries relating to seminal studies conducted in an academic environment, university or private research institute, where the process of filing patent applications is, in most cases, dictated by the publication process of the scientific manuscripts relating these inventions. Nonetheless, several patents filed originally by universities or private research institute, and referenced in this manuscript, have originated start-ups, giving them a research niche and value for raising funds. Successful pharmaceutical/biotech companies have built-up their portfolios of intellectual property from these 'early' patents, giving other protections to their discoveries and securing more time in their endeavour to bring their technology to therapy.

In all, adult-derived neural progenitor and stem cells represent a tremendous opportunity for cellular and gene therapy. But, their potential to restore brain function remains to be validated. In this respect, limitations, pitfalls and risks need to be assessed and resolved. Among these, NSCs need to be identified unequivocally and characterized, neural progenitor and stem cells need to be maintained as homogenous populations in culture, and their broader potentials need to be fully understood and characterized. Their tumorigenic and inflammatory potentials must be monitored. Gene-based therapy of NSCs may also bolster cellular therapy, broadening their therapeutic potential to genetic diseases. Beside its therapeutic potential, NSC biology has the potential to better understand and unravel the mechanisms underlying neurological diseases and disorders, particularly by isolating and characterizing neural progenitor and stem cells from patients, leading to new discoveries and clinical applications.

Declaration of Interest

The author states no conflict of interest and has received no payment in preparation of this manuscript.

Acknowledgments

Reproduced with permission from Informa Pharmaceutical Science: Taupin P. Therapeutic neuronal stem cells: patents at the forefront. Expert Opinion on Therapeutic Patents 18(10):1107-10.

References

Papers of special note have been highlighted as either of interest (•) or of considerable interest (••) to readers.

[1] McKay R. Stem cells in the central nervous system. *Science* 1997; 276: 66-71

[2] Eriksson PS, Perfilieva E, Bjork-Eriksson T, et al. Neurogenesis in the adult human hippocampus. *Nat. Med.* 1998; 4: 1313-7
•• Tissue samples originate from patients treated with bromodeoxyuridine (BrdU), a marker of DNA synthesis, as part of their treatments. Immunohistology for BrdU reveals the presence of newborn neuronal cells, born during the treatment, with BrdU, in the adult DG.

[3] Curtis MA, Kam M, Nannmark U, et al. Human neuroblasts migrate to the olfactory bulb via a lateral ventricular extension. *Science* 2007; 315: 1243-9

[4] Taupin P. Neural progenitor and stem cells in the adult central nervous system. *Ann. Acad. Med. Singapore* 2006; 35: 814-7

[5] Weiss S, Reynolds B. Novel growth factor-responsive progenitor cells which can be proliferated in vitro. WO/1993/001275; 1993

[6] Weiss S, Reynolds BA, Van Der Kooy D, et al. In situ modification and manipulation of stem cells of the central nervous system. WO/1995/013364; 1995

[7] Weiss S, Reynolds B. In vitro induction of dopaminergic cells. WO/1996/015224; 1996

[8] Weiss S, Reynolds B. Regulation of neural stem cell proliferation. WO/1996/015226; 1996

[9] Reynolds BA, Weiss S. Generation of neurons and astrocytes from isolated cells of the adult mammalian central nervous system. *Science* 1992; 255: 1707-10

•• Neural progenitor and stem cells are isolated from the striatal area containing the SVZ, one of the main neurogenic regions of the adult mammalian brain. Neural progenitor and stem cells are propagated and cultured, as neurospheres, in defined medium, in the presence of high concentration of trophic factor, epidermal growth factor.

[10] Reynolds BA, Tetzlaff W, Weiss S. A multipotent EGF-responsive striatal embryonic progenitor cell produces neurons and astrocytes. *J. Neurosci.* 1992; 12: 4565-74

[11] Palmer TD, Schwartz PH, Taupin P, et al. Cell culture. Progenitor cells from human brain after death. *Nature* 2001 ; 411: 42-3

• Post-mortem neural progenitor and stem cells were isolated and cultured in vitro, in defined medium, in the presence of a cocktail of trophic factors and CCg.

[12] Palmer TD, Gage FH, Schwartz PH, Taupin PJ. Postmortem stem cells. WO/2002/036749; 2002

[13] Uchida N. Use of collagenase in the preparation of neural stem cell cultures. WO/2000/050572; 2000

[14] Gage FH, Taupin PJ, Ray. Co-factors for trophic factors, and methods of use thereof. JWO/2000/033791; 2000

[15] Taupin P, Ray J, Fischer WH, et al. FGF-2-responsive neural stem cell proliferation requires CCg, a novel autocrine/paracrine cofactor. *Neuron* 2000; 28: 385-97

[16] Bjornson CR, Rietze RL, Reynolds BA, et al. Turning brain into blood: a hematopoietic fate adopted by adult neural stem cells in vivo. *Science* 1999; 283: 534-7

[17] Brazelton TR, Rossi FM, Keshet GI, et al. From marrow to brain: expression of neuronal phenotypes in adult mice. *Science* 2000; 290: 1775 -9

[18] Mezey E, Chandross KJ, Harta G, et al. Turning blood into brain: cells bearing neuronal antigens generated in vivo from bone marrow. *Science* 2000; 290: 1779-82

[19] Toma JG, Akhavan M, Fernandes KJ, et al. Isolation of multipotent adult stem cells from the dermis of mammalian skin. *Nat. Cell Biol.* 2001; 3: 778-84

[20] D'Amour KA, Gage FH. Are somatic stem cells pluripotent or lineage-restricted? *Nat. Med.* 2002; 8: 213 -4

[21] Bjornson CR, Rietze RL, Reynolds BA, Vescovi AL. Generation of hematopoietic cells from multipotent neural stem cells. WO/1999/016863; 1999

[22] Lagasse E, Weissman IL. Pluripotential stem cells. WO/2001/071016; 2001

[23] Takahashi K, Yamanaka S. Induction of pluripotent stem cells from mouse embryonic and adult fibroblast cultures by defined factors. *Cell* 2006; 126: 663-76

[24] Yu J, Vodyanik MA, Smuga-Otto K, et al. Induced pluripotent stem cell lines derived from human somatic cells. *Science* 2007; 318: 1917-20

[25] Gage FH, Coates PW, Palmer TD, et al. Survival and differentiation of adult neuronal progenitor cells transplanted to the adult brain. *Proc. Natl. Acad. Sci. USA* 1995; 92: 11879 -83

[26] Weiss S, Reynolds B, Hammang JP, Baetge EE. In vitro growth and proliferation of genetically modifi ed multipotent neural stem cells and their progeny. US-05,750,376; 1998

[27] Shihabuddin LS, Numan S, Huff MR, et al. Intracerebral transplantation of adult mouse neural progenitor cells into the Niemann-Pick-A mouse leads to a marked decrease in lysosomal storage pathology. *J. Neurosci.* 2004; 24: 10642-51

[28] Uchida N, Jacobs Y, Tamaki S. Methods for the treatment of lysosomal storage disorders. WO/2006/074387; 2006

[29] Taupin P. HuCNS-SC (StemCells). *Curr. Opin. Mol. Ther.* 2006; 8: 156-63

[30] StemCells, Inc. website. Available from: http://www.stemcellsinc.com

[31] NeuroGeneration website. Available from: http://www.neurogeneration.com

[32] Seaberg RM, van der Kooy D. Adult rodent neurogenic regions: the ventricular subependyma contains neural stem cells, but the dentate gyrus contains restricted progenitors. *J. Neurosci.* 2002; 22: 1784 -93

[33] Ray J, Gage FH. Differential properties of adult rat and mouse brain-derived neural stem/progenitor cells. *Mol. Cell Neurosci.* 2006; 31: 560-73

[34] Reynolds BA, Rietze RL. Neural stem cells and neurospheres-re-evaluating the relationship. *Nat. Methods* 2005; 2: 333-6

[35] Buck DW, Uchida N, Weissman I. Enriched central nervous system cell populations. WO/2000/047762; 2000

[36] Uchida N, Capela A. Enriched central nervous system stem cells and progenitor cell populations, and methods for identifying, isolating and enriching for such populations. WO/2004/020597; 2004

[37] Uchida N, Capela A. Enriched central nervous system stem cell and progenitor cell populations, and methods for identifying, isolating and enriching for such populations. US-7,381,561; 2008

Chapter X

Adult Neural Stem Cells for the Treatment of Neuroinflammation

Abstract

Background: The application is in the field of stem cells and their therapeutic application for the treatment of inflammation. *Objective:* It aims at characterizing the potential of adult-derived neural progenitor and stem cells for the treatment of inflammation of the central nervous system (CNS). *Methods:* Neural progenitor and stem cells were isolated and expanded from the subventricular zone (SVZ) of adult mice (aNSCs). They were administered intravenously in an animal model of multiple sclerosis (MS). *Results:* Mice transplanted either at the disease onset or at the onset of the first relapse show clinical signs of improvements. Adult NSCs exert their therapeutic activity by reducing neuroinflammation. *Conclusion:* The application claims the use of aNSCs and multipotent somatic stem cells for the treatment of inflammation, associated with neurological diseases, disorders and injuries particularly, and for inducing tolerance to the immune central and/or peripheral system.

1. Introduction

Somatic stem cells are self-renewing cells that generate the main cell types of the body [1]. Embryonic stem cells are pluripotents. They generate cell types of the three germ layers of the embryos: the neurectoderm, mesoderm and endoderm. Embryonic stem cells are derived from the inner cell mass of blastocysts [2]. Fetal and adult stem cells are multipotents, generating the cell types from the tissue from which they originate. During development, stem cells contribute to the formation of the tissues. In the adult, stem cells contribute to homeostasis of the tissues and regeneration after injury. Stem cells can be isolated from various adult tissues, including the bone marrow, liver and skin [3]. The adult brain used to be believed to be devoid of stem cells. Underlying this belief was the fact that the adult brain lacks the ability to self-repair, particularly in neurodegenerative diseases, like Alzheimer's and Parkinson's diseases, and after injury. Studies in the 1980s and 1990s reported celi proliferation and neurogenesis in two regions of the adult brain, the SVZ and the dentate gyrus of the hippocampus, in rodents [4]. In 1992, Reynolds and Weiss reported the isolation and characterization of neural progenitor and stem cells from the striatal area, including the SVZ, of adult mice brains [5]. Neural stem cells (NSCs) are the self-renewing multipotent cells that generate the main phenotypes of the nervous system: neurons, astrocytes and oligodendrocytes. Neural progenitor and stem cells have been isolated and characterized from various regions of the adult CNS, including the spinal cord, and from various species, including human post-mortem tissues [4]. Neurogenesis has been reported in the dentate gyrus and SVZ of various mammalian species, including humans [6,7]. The confirmation that adult neurogenesis occurs in the adult brain and that NSCs reside in the adult CNS, opens new opportunities for the treatment of neurological diseases and injuries, particularly neurodegenerative diseases. It is proposed to stimulate endogenous neural progenitor or stem cells or to transplant adult-derived neural progenitor and stem cells, to repair the degenerated or injured pathways [8,9].

Inflammation is a self-defensive process by which cells of the immune system and pro-inflammatory substances of the body protect us from infections, foreign substances and injuries. During the inflammatory response, the immune cells release a host of powerful regulatory substances, including glutamate, nitric oxide, reactive oxygen species, cytokines, chemokines, interferons (IFNs) and interleukins (ILs) [10,11]. These chemicals produce both beneficial and harmful effects to the cellular environment, creating further damage. In the CNS, inflammatory responses occur in a broad range of neurological diseases, disorders and injuries, including Alzheimer's disease, cerebral strokes, traumatic brain and spinal cord injuries [12,13]. Chronic inflammation may also be a causative factor for neurological diseases and disorders and particularly for neurodegenerative diseases, like Alzheimer's disease [14]. Multiple sclerosis (MS) is an autoimmune disease of the nervous system that leads to paralysis and disabilities. It is associated with loss of axons and myelin sheaths [15] and is characterized by delayed-type hypersensitivity (Th1) inflammatory responses. Experimental autoimmune encephalomyelitis (EAE) is an animal model of MS. In MS and EAE, Th1 cells and the expression of Th1-associated cytokines (IFN-γ and IL-6) and of Th1-inducing cytokine (IL-12) is associated with active or relapse of the disease, whereas Th2 cells and the expression of the Th2 cytokines, IL-4, IL-10 and IL-13, are associated with remission or suppression of the disease [16].

Neural progenitor and stem cells, administered intravenously, migrate to diseased and injured sites of the brain [17]. When administered in the EAE model of MS, adult NSCs (aNSCs) migrate to the sites of inflammation in the CNS and induce functional recovery [18]. The application aims to characterize the therapeutic potential of aNSCs for the treatment of MS particularly and neuroinflammatory diseases in general. It aims to identify the cellular and molecular mechanisms underlying the therapeutic activity of transplanted aNSCs in the EAE model.

2. Chemistry

Cultures of neural progenitor and stem cells (aNSCs) were established from the SVZ of adult (6- to 8-week-old) SJL and C57BL/6 mice, based on the protocol originally described by Reynolds and Weiss (in 1992) and modified by Pluchino *et al.* (in 2003) [5,18]. Briefly, the brains were isolated and the SVZ regions dissected out. During this process, the subependyma was removed from the dissected tissue. The tissue was cut into 1-mm 3 pieces and digested with trypsin, hyaluronidase and kynurenic acid. The dissociation of the tissue was completed by mechanical trituration, with fire-polished Pasteur pipettes. Cells were resuspended and cultured in DMEM/F12 medium, containing 2 mM L-glutamine, 0.6% glucose, 9.6 mg/ml putrescine, 6.3 ng/ml progesterone, 5.2 ng/ml sodium selenite, 0.025 mg/ml insulin, 0.1 mg/ml transferrin and 2 μg/ml heparin. Cells were cultured in the presence of NS-A medium (Euroclone) containing 20 ng/ml of epidermal growth factor and 10 ng/ml basic fibroblast growth factor. Adult neural progenitor and stem cells were cultured, as neuropheres, in untreated tissue culture flasks. Cells under 15 passages were used.

For transplantation studies, aNSCs were labeled *in vitro*, with *Escherichia coli* -derived β-galactosidase (LacZ), using a third-generation lentiviral vector pRRLsin.PPT-hCMV, as previously described [18].

3. Biology and Action

3.1. Relapsing-remitting Model of Experimental Autoimmune Encephalomyelitis

The relapsing-remitting model of EAE (R-EAE) was used. R-EAE was induced experimentally by administration of 200 μg PLP139-151 in complete Freund's adjuvant in SJL mice [19].

3.2. Syngeneic Transplantation

β-gal-labelled aNSCs were dissociated as single cells and injected intravenously into R-EAE mice through the tail vein: 1 - 2.10 6 cells in 150 μl, prepared in PBS, were administered per mouse. The cells were administered at the first episode of the disease [13.1 ± 0.3 days post immunization (dpi)] or at the occurrence of first clinical relapse (30.9 ± 1.1 dpi). Clinical relapse is assessed by recording daily the body weight and clinical score (0 = healthy, 1 = limp tail, 2 = ataxia and/or paresis of hind limbs, 3 = paralysis of hind limbs and/or paresis of forelimbs, 4 = tetra paralysis, 5 = moribund or death). Clinical relapses are defined as the occurrence of 0.5 increase of the clinical score persisting for a minimum of three consecutive days [19]. Mice were followed up to three months post-transplantation, after which they were sacrificed.

3.3. In Situ Proliferation of Transplanted aNSCs

Three days before sacrifice, R-EAE mice were given one dose of bromodeoxyuridine (BrdU, 50 mg/kg) per day. BrdU is a thymidine analog used for birth dating and monitoring cell proliferation.

3.4. Immunocytology and Immunohistology

For immunocytology, neurospheres were cultured onto glass chamber slides coated with matrigel. For immunohistology, brain tissues were embedded in paraffin or in agarose, and were processed according to standard procedures. In particular, brain sections were stained with Luxol fast blue and Bielshowwsky for detecting inflammatory inflitrates, demyelination and axonal loss [20]. The cells and tissues were processed for immunofluorescence, applying standard procedures. A broad range of primary antibodies was used, including antibodies againts BrdU, nestin, neuronal class III β-tubulin, NG2, CD45, CD31 and FasL/CD95-ligand.

A range of assays and studies were conducted, both *in vitro* and *in vivo*, to assess the cellular and molecular mechanisms underlying the therapeutic activity of transplanted aNSCs in the EAE model. This includes adhesion assays, chemotaxis assays, cytokine and chemokine assays, apotosis experiments, RT-PCR, flow cytometry, FACS analysis and intravital microscopy.

The results show that R-EAE mice, administered with aNSCs, elicit signs of clinical improvements and that aNSCs induce the programmed cell death of immune cells involved in pro-inflammatory response. Clinical improvements were observed whether aNSCs were administered at the disease onset or at the onset of the first relapse. Mice transplanted at disease onset recover between 30 and 60 dpi. During this period, they develop 2-fold fewer clinical relapses than sham-treated mice. Mice transplanted at disease onset maintained a low rate of relapse than sham-treated mice, up to the end of the follow-up study. Mice transplanted at the onset of the first relapse starts to recover later, but elicit a 3-fold reduction of the relapse rate between 60 and 90 dpi, when compared to sham-treated mice. R-EAE mice transplanted with aNSCs are characterized by a pattern of programmed cell death of pro-inflammatory Thl cells, and not of the anti-inflammatory Th2 cells, in the inflamed perivascular areas of the CNS. Results further show that transplanted aNSCs survive the inflammatory episodes and maintain a pool of undifferentiated cells with the ability to proliferate, in the inflammed areas, up to the end of the follow-up study. This shows that aNSCs exert an anti-inflammatory activity, in the R-EAE model of MS, by reducing the pro-inflammatory response in the CNS. The transplanted aNSCs migrate to the inflamed areas of the CNS, where they maintain a pool of cells. aNSCs may retain the ability to protect against recurrent episodes of inflammation; they also promote endogenous myelin-producing cells to acquire a mature phenotype and replace damaged neural cells. The neuroprotective activity of aNSCs during neuroinflammation in the R-EAE model reveals a therapeutic potential for aNSCs for the treatment of inflammation and inflammatory diseases of the nervous system [21].

4. Expert Opinion

The present application claims the use of adult-derived neural progenitor and stem cells for the treatment of neuroinflammation and inflammatory diseases, particularly of the nervous system, and for inducing tolerance to the immune central and/or peripheral system.

The transplanted adult-derived neural progenitor and stem cells prevent or decrease inflammation through either the induction of programmed cell death (apoptosis) of blood-borne CNS-infiltrating pro-inflammatory Th1 cells or/and through an immunomodulatory mechanism leading to immune tolerance. The transplanted stem cells maintain their ability to protect the CNS from chronic inflammatory reactions. The anti-inflammatory property of adult-derived neural progenitor and stem cells may be used for the treatment of inflammatory diseases, and also for the treatment of a broad range of neurological diseases, disorders and injuries in which neuroinflammation is involved, like Alzheimer's disease, cerebral strokes, depression, epilepsy, Parkinson's disease, and traumatic brain and spinal cord injuries [14,15]. It can also be used to treat other inflammatory diseases or inflammation associated with other diseases, like diabetes and rheumatoid arthritis. A strength of the present application is that the administration of the stem cells, by intravenous injection, is a non-invasive procedure to treat neuroinflammation.

Stem cells reside in specialized microenvironments or 'niches' [22]. Such niches regulate and control the self-renewal and differentiation activities of NSCs. An angiogenic and astroglial niche for neurogenesis has been identified and characterized in the adult brain [23]. It is proposed that the perivascular areas, sites of neuroinflammation in the R-EAE mice, would create a microenvironment where neural progenitor and stem cells from the bloodstream would migrate and accumulate. These 'atypical perivascular niches' in the CNS would control the development and fate of the adult-derived neural progenitor and stem cells, and therefore their therapeutic potential for the treatment of neuroinflammatory responses. This includes the maintenance of a pool of undifferentiated neural progenitor and stem cells, to protect the CNS from chronic inflammatory reactions. The migration of the adult-derived neural progenitor and stem cells to the sites of inflammation in the CNS would be facilitated by the disruption of the blood–brain barrier, as a consequence of the neuroinflammation process [24]. This disruption allows cells from the haematopoietic system and from the administered neural progenitor and stem cells to leave the bloodstream and come into contact with the site of injury, to exert their biological activities.

The potential of adult-derived neural progenitor and stem cells to reduce neuroinflammation broadens the therapeutic potential of adult-derived neural progenitor and stem cells to the treatment of inflammation and inflammatory diseases. It not only offers new opportunities and avenues for the treatment of inflammation and inflammatory diseases, but also neurological diseases, disorders and injuries, by attacking one of their causes. There are, however, questions that remain to be answered. Among them, additional controls might be needed to confirm the genetic labelling of the transplanted aNSCs *in situ*. What would be the optimal source of stem cells to use for such therapy, given that autologous and syngeneic transplantations are not likely candidates for the transplantation of adult-derived neural progenitor and stem cells? And how could such strategy applied to reduce neuroinflammation in allogeneic transplantations, particularly in the nervous system, to reduce the risk of tissue rejection?

Declaration of Interest

The author states no conflict of interest and has received no payment in preparation of this manuscript.

Acknowledgments

Reproduced with permission from Informa Pharmaceutical Science: Taupin P. Adult neural stem cells for the treatment of neuroinflammation. Expert Opinion on Therapeutic Patents 19(3): 373-6. Copyright 2009, Informa UK Ltd.

Patent details

Title: Inflammation
Assignee: Fondazione Centro San Raffaele del Monte Tabor
Inventors: Pluchino S, Martino G.
Priority data: 12/07/2005
Filling date: 12/07/2006
Publication date: 08/02/2007
Publication no: WO2007015173

References

Papers of special note have been highlighted as either of interest (•) or of considerable interest (••) to readers.

[1] Potten CS, Loeffler M. Stem cells: attributes, cycles, spirals, pitfalls and uncertainties. Lessons for and from the crypt. *Development* 1990; 110: 1001-20
• Review defi ning stem cells criteria.

[2] Taupin P. Derivation of embryonic stem cells for cellular therapy: Challenges and new strategies. *Med. Sci. Monit.* 2006; 12: RA75-78

[3] Stocum DL, Zupanc GK. Stretching the limits: Stem cells in regeneration science. *Dev. Dyn.* 2008; 237: 3648-71

[4] Taupin P, Gage FH. Adult neurogenesis and neural stem cells of the central nervous system in mammals. *J. Neurosci. Res.* 2002; 69: 745-9

[5] Reynolds BA, Weiss S. Generation of neurons and astrocytes from isolated cells of the adult mammalian central nervous system. *Science* 1992; 255: 1707-10

•• Original report of the isolation and characterization of neural progenitor and stem cells from the adult brain of mammals.

[6] Eriksson PS, Perfilieva E, Bjork-Eriksson T, et al. Neurogenesis in the adult human hippocampus. *Nat. Med.* 1998; 4: 1313-7

• First evidence of adult neurogenesis in the human brain.

[7] Curtis MA, Kam M, Nannmark U, et al. Human neuroblasts migrate to the olfactory bulb via a lateral ventricular extension. *Science* 2007; 315: 1243-9

[8] Taupin P. The therapeutic potential of adult neural stem cells. *Curr. Opin. Mol. Ther.* 2006; 8: 225-31

[9] Taupin P. Therapeutic neuronal stem cells: patents at the forefront. *Exp. Opin. Ther. Patents* 2008; 18: 1107-10

[10] 10. Jander S, Schroeter M, Stoll G. Interleukin-18 expression after focal ischemia of the rat brain: association with the late-stage inflammatory response. *J. Cereb. Blood Flow Metab.* 2002; 22: 62-70

[11] Ghirnikar RS, Lee YL, Eng LF. Inflammation in traumatic brain injury: role of cytokines and chemokines. *Neurochem. Res.* 1998; 23: 329-40

[12] Nencini P, Sarti C, Innocenti R, et al. Acute inflammatory events and ischemic stroke subtypes. *Cerebrovasc. Dis.* 2003; 15: 215-21

[13] Schmidt OI, Heyde CE, Ertel W, et al. Closed head injury – an inflammatory disease? *Brain Res. Brain Res. Rev.* 2005; 48: 388-99

[14] Arnaud L, Robakis NK, Figueiredo-Pereira ME. It may take inflammation, phosphorylation and ubiquitination to 'tangle' in Alzheimer's disease. *Neurodegener. Dis.* 2006; 3: 313-9

[15] Minghetti L. Role of inflammation in neurodegenerative diseases. *Curr. Opin. Neurol.* 2005; 18: 315-21

• Inflammation as a causative factor of neurological diseases and disorders.

[16] Owens T, Wekerle H, Antel J. Genetic models for CNS inflammation. *Nat. Med.* 2001; 7: 161-6

[17] Brown AB, Yang W, Schmidt NO, et al. intravascular delivery of neural stem cell lines to target intracranial and extracranial tumors of neural and non-neural origin. *Hum. Gene Ther.* 2003; 14: 1777-85

[18] Pluchino S, Quattrini A, Brambilla E, et al. Injection of adult neurospheres induces recovery in a chronic model of multiple sclerosis. *Nature* 2003; 422: 688-94

•• Original report for the therapeutic potential of adult-derived neural progenitor and stem cells for the treatment of neuroinflammation.

[19] McRae BL, Kennedy MK, Tan LJ, et al. Induction of active and adoptive relapsing experimental autoimmune encephalomyelitis (EAE) using an encephalitogenic epitope of proteolipid protein. *J. Neuroimmunol.* 1992; 38: 229-40

[20] Lindvall O, Kokaia Z, Martinez-Serrano A: Stem cell therapy for human neurodegenerative disorders-how to make it work. *Nat. Med.* 2004; 10 (Suppl): S42-50

[21] Pluchino S, Zanotti L, Rossi B, et al. Neurosphere-derived multipotent precursors promote neuroprotection by an immunomodulatory mechanism. *Nature* 2005; 436: 266-71

[22] Mitsiadis TA, Barrandon O, Rochat A, et al. Stem cell niches in mammals. *Exp. Cell Res*. 2007; 313: 3377-85

[23] Ninkovic J, Götz M. Signaling in adult neurogenesis: from stem cell niche to neuronal networks. *Curr. Opin. Neurobiol*. 2007; 17: 338-44

[24] Lossinsky AS, Shivers RR. Structural pathways for macromolecular and cellular transport across the blood-brain barrier during infl ammatory conditions. *Histol. Histopathol.* 2004; 19: 535-64

Chapter XI

Magnetic Resonance Imaging for Monitoring Neurogenesis in the Adult Hippocampus

Abstract

Background: The application is in the field of bioimaging and adult neurogenesis. Objective: It aims at correlating the volume of cerebral blood (CBV) in the dentate gyrus (DG) of the human hippocampus, determined by magnetic resonance imaging (MRI), with neurogenesis in the brain of adult rodents. Methods: Adult mice were submitted to voluntary exercise or administration of fluoxetine or valproic acid (VPA). The CBV of DG was determined by MRI and neurogenesis was quantified by immunohistofluorescence. The CBV in human subjects selected and grouped according to their fitness activity was determined by MRI in the DG. Results: A selective increase in the CBV of the DG is observed in rodents housed in activity cages or administered with fluoxetine and VPA. A selective increase in the CBV of the DG is also observed in exercising humans. The selective increase of the CBV in the human DG correlates with the selective increase of the CBV in the DG and neurogenesis induced by exercise or fluoxetine and VPA in rodents. Conclusion: This indicates that neurogenesis is increased in the DG of exercising humans. The application claims the imaging of the DG of patients by MRI as a paradigm to monitor neurogenesis and identify treatments involving stimulation of neurogenesis.

1. Introduction

It is now accepted that neurogenesis occurs in the adult brain and neural stem cells reside in the adult central nervous system of mammals [1,2]. Neurogenesis occurs throughout adulthood in the dentate gyrus (DG) of the hippocampus and subventricular zone along the ventricles, in various species including humans [3,4]. In the DG, newly generated neuronal cells in the subgranular zone migrate to the granule cells layer, where they differentiate and extend axonal projections to the CA3 region, like mature granule cells [5,6]. Newly generated granule-like cells establish functional connections in the CA3 region and survive for an extended period of time [3,7]. Neurogenesis in the adult brain, particularly the hippocampus, is modulated by a broad range of environmental stimuli, physiological and pathological processes, trophic factors/cytokines and drugs [8]. Neurogenesis decreases with age and is diminished by chronic stress and depletion of serotonin, whereas an enriched environment, hippocampal-dependent learning tasks, and voluntary exercise increase neurogenesis in the adult hippocampus of rodents and non-human primates [9-15]. Neurogenesis in the adult hippocampus is involved in the formation of trace memories, a hippocampal-dependent form of learning and memory, in rodents [16]. In humans, neurogenesis is enhanced in the hippocampus of patients with Alzheimer's disease (AD) [17]. The adult brain has therefore the potential of self-repair, and newly generated neuronal cells in the adult brain may play a critical role in physiological and pathological processes, such as learning and memory, AD and depression [18].

Chronic antidepressant treatments, such as fluoxetine treatments, valproic acid (VPA), a drug used in the treatment of long-term epilepsy, and drugs currently used to treat patients with AD, such as galantamine and memantine, increase neurogenesis in the adult hippocampus [19-21]. X-ray irradiation of the hippocampal region inhibits neurogenesis in the DG and prevents the behavioural effect of antidepressants, such as fluoxetine, in adult mice [22]. Hence, in addition to their therapeutic potential for cellular therapy, adult neurogenesis may also play a critical role in the pharmacology of neurological diseases and disorders, such as learning and memory impairments, AD, depression and epilepsy [23-25]. Alzheimer's disease is a neurodegenerative disease associated with the loss of nerve cells in areas of the brain that are vital to memory and other mental abilities, such as the entorhinal cortex and hippocampus [26,27]. Depression is associated with reduction in the volume of the hippocampus in patients with major depression, suggesting the loss of nerve cells in the hippocampus of those patients [28-30]. Hence, there is a significant interest in identifying and characterising conditions, drugs and molecules that stimulate neurogenesis in the adult hippocampus of humans in an attempt to compensate or reverse neuronal loss, alleviate symptoms associated with learning and memory deficits, AD, depression and epilepsy, and treat these diseases and disorders.

There are numerous protocols and paradigms used to study adult neurogenesis. The most used at present is the bromodeoxyuridine (BrdU)-labelling paradigm. BrdU is a thymidine analogue used for birth dating and monitoring cell proliferation [31,32]. It consists of administering BrdU and processing, staining and analysing the tissues for BrdU and other markers of interest, by immunohistology and confocal microscopy [33-36]. This paradigm is not without pitfalls and limitations and is not applicable for human studies, with the exception of rare cases in which patients were treated with BrdU as part of their cancer therapy and donated tissue samples for research investigations [3,4]. Other modes of investigation involve retroviral labelling and transgenic animal models and are also limited to animal studies. In humans, post-mortem immunohistology is most commonly used to investigate cell proliferation and adult neurogenesis [17]. Another technique involving birth dating using ^{14}C has been used successfully to investigate adult neurogenesis in humans [37]. This latter technique allows retrospective studies in post-mortem samples. MRI is a medical imaging technique for visualising the structure and function of the body. It uses a powerful magnetic field and non-ionising radiation to align the nuclear magnetisation of usually hydrogen atoms in water in the body. It provides detection with high sensitivity of the volume of cerebral blood (CBV) [38]. Agents such as gadolinium and Omniscan (gadodiamide) are used to generate image contrast to show adequately the anatomy or pathology of interest. MRI is a very sensitive, safe, non-invasive procedure that provides real time detailed images of the living body in any plane. It has sufficient spatial resolution to visualise regions and subregions of the brain, such as the hippocampus and DG, in humans and rodents [39,40]. MRI techniques have recently been developed and applied to study newly generated neuronal cells and transplanted neural progenitor and stem cells in the adult brain of rodents [41-43].

The application aims at devising an MRI paradigm to study adult neurogenesis in the hippocampus and its modulation in rodents and humans [44]. The application may be used to study adult neurogenesis in the hippocampus of humans in real time and its modulation, by various conditions and drug treatments.

2. Chemistry

2.1. Experimental Study in Rats

Male F344 rats age 6 – 8 weeks (150 – 250 g) were used in this study. Rats were housed individually and divided into control and test groups. Group I or control group: animals were housed in standard laboratory cages. Group II: animals were housed in activity cages, with a running wheel linked to a computer monitoring the wheel's use. Groups III and IV: animals were housed in standard laboratory cages and were administered drugs known to stimulate neurogenesis in the adult hippocampus, fluoxetine and VPA, respectively. Animals received daily oral injections of 10 mg/kg fluoxetine for 28 days (Group III). Animals received 2 daily intraperiteoneal injections of 300 mg/kg VPA for 28 days. VPA was also provided in the drinking water (12 g/l) (Group IV). There was a minimum of 12 animals per group. All animals received 1 daily intraperiteoneal injection of 100 mg/kg BrdU for 7 consecutive days beginning the first day of experiment/drug treatment (day 1). All animals were analysed by MRI for determination of CBV at day 1 and day 28. All mice were imaged with an MRI protocol to estimate the CBV in the entorhinal cortex, the DG, the CA3 and CA1 regions and the subiculum [39]. At the end of the experimental study, the animals were killed. Their brains were extracted and processed for histology to study and quantify neurogenesis.

2.2. MRI Scanning in Rats

MRI scans in rats were performed using a Bruker AVANCE 400WB spectrometer (Bruker NMR, Inc., Bilerica, MA, US) with an 89-mm-bore 9.4 T vertical Bruker magnet (Oxford Instruments Ltd, UK), a birdcage RF probe and a shielded gradient system up to 100 G/cm. Gadolinium (gadodiamide) sterile aqueous solution at a concentration of 287 mg/ml pH 5.5 – 7.0 is injected undiluted by means of a catheter, with an outer diameter of 0.6 mm.

The catheter is placed intraperitoneally before imaging and secured with 6.0 silk suture materials.

2.3. Histology

Histology for BrdU and neuronal markers was performed according to previously published protocols and procedures [9,10,14,45].

2.4. Epidemiological Study

Twenty individuals, 20 – 45 years of age, non-smokers, sedentary and habitual non-exercisers, who qualify as below average fitness by the standards of the American Heart Association, were recruited for this study. Subjects were divided in two groups. Group I: subjects were submitted to exercise with moderate intensity training on a selection of aerobic activities, such as cycling on a stationary ergometer, running on a treadmill, climbing on a stairmaster or using an elliptical trainer. Subjects in this group exercised at a heart rate of 55 – 65% of their maximum heart rate for 2 weeks and 65% of maximum heart rate for 10 weeks. Group II: subjects were submitted to exercise with high-intensity training on a selection of aerobic activities, such as cycling on a stationary ergometer, running on a treadmill, climbing on a stairmaster or using an elliptical trainer. Subjects in this group exercised at a heart rate of 55 – 65% of their maximum heart rate for 2 weeks, 65 – 75% of their maximum heart rate for the next 2 weeks and 75% of maximum heart rate for 8 weeks. Cardiovascular indices and respiration were monitored and recorded. All subjects were imaged with two MRI scans, the first at the start of the study and the second at the end of the study. All subjects were imaged with an MRI protocol to estimate the CBV in the four hippocampal subregions: the entorhinal cortex, the DG, the CAl region and the subiculum [38]. The inter-individual differences in physical activity were measured by determining the maximum volume of oxygen consumption, to quantify individual differences in degree of exercise [46].

2.5. MRI Protocol in Human Subjects

Two sets of coronal three-dimensional Tl-weighted images were acquired, one before and the second 4 min after intravenous administration of Omniscan (0.1 mmol/kg). Images were acquired perpendicular to the long axis of the hippocampus and processed using analysis software packages (MEDx Sensor Systems). The hippocampus and the four hippocampal subregions, the entorhinal cortex, the DG, the CAl region and the subiculum, were identified as follows. Hippocampus: the hippocampus is identified by following the trace of the hippocampal sulcus and of the internal white matter tracts. Entorhinal cortex: the lateral and inferior boundary of the entorhinal cortex follows the collateral sulcus, the medial boundary is the medial aspect of the temporal lobe and the superior boundary is delineated by the hippocampal sulcus and the grey/white distinction between subiculum and entorhinal cortex. Dentate gyrus: the medial boundary of the DG is the medial extent of the temporal lobe, the inferior/lateral boundary is the hippocampal sulcus/white matter tracts and the superior boundary is the top of the hippocampal formation, where the alveus is typically identified. CA1 region: the medial boundary of the CA1 region is 2 – 3 pixels lateral to the end of the subiculum, approximately at the beginning of the vertical inflection of the hippocampus and of the extension of the hippocampal sulcus/white matter tracts, the inferior boundary is the white matter of the underlying parahippocampal gyrus, and the superior boundary is the top of the hippocampal formation. Subiculum: the medial boundary of the subiculum is the medial extent of the hippocampal sulcus and/or the horizontal inflection of the hippocampus, the inferior boundary is the white matter of the underlying parahippocampal gyrus, the superior boundary is the hippocampal sulcus and the lateral boundary is a few pixels medial to the vertical inflection of the hippocampus.

3. Biology and Action

The selective increase of the CBV in the DG is determined by subtracting the curve of the average CBV of the other hippocampal subregions from the curve of the CBV generated from the DG by MRI, in animal and human studies.

These studies reveal that a selective increase in the CBV of the DG is observed in rodents housed in activity cages with a running wheel, or administered drugs known to stimulate neurogenesis in the adult hippocampus, fluoxetine and VPA. A selective increase in the CBV of the DG is also observed in exercising humans. The selective increase of the CBV in the human DG correlates with the selective increase of the CBV in the DG and neurogenesis induced by exercise or modulated with the drugs fluoxetine and VPA, in the rodent study. This indicates that neurogenesis is increased in the DG of exercising humans.

4. Expert Opinion

The application provides a paradigm to indicate safely and in real time by MRI whether a condition or drug increases neurogenesis in the DG of living mammals, including humans. To this end, the CBV of the DG and other hippocampal subregions, such as the entorhinal cortex, the DG, the CAl region and the subiculum, are determined by MRI. In rodents, the selective increase of the CBV in the DG, determined by subtracting the curve of the average CBV of the other hippocampal subregions from the curve of the CBV generated from the DG, is correlated with increased neurogenesis in the subjects. In humans, the selective increase of the CBV in the DG is correlated with the selective increase of the CBV in the DG with control subjects and with the selective increase of the CBV in the DG and increased neurogenesis in animal studies. Based on this paradigm, the results presented indicate that neurogenesis is increased in the DG of exercising humans. This application claims the measure of the selective increase of the CBV in the DG as a paradigm to study adult neurogenesis and its modulation, by various conditions and drugs, in humans.

The approach is based on the tight spatial and temporal coupling between neurogenesis and angiogenesis. Neurogenesis in the adult brain, particularly the hippocampus, occurs at sites of angiogenesis, defining the angiogenic niche for neurogenesis [47-51]. Hence, angiognesis reflects neurogenesis in the adult DG. Angiogenesis results in an increase in the CBV regionally. Measure of the CBV by MRI has been applied to monitoring angiogenesis [52]. MRI has sufficient spatial resolution to visualise subregions of the hippocampus, such as the entorhinal cortex, the DG, the CAl region and the subiculum, in humans and rodents [39]. The measure of the CBV and its variation regionally by MRI are therefore indicative of neurogenesis and its modulation in the adult DG.

Other factors and parameters, such as cardiac output and synaptic activity, affect regionally the CBV in the brain. Hence, measure of the CBV and its variation regionally by MRI are not specifically associated with neurogenesis. To address this issue, the authors normalised the data by subtracting the curve of the average CBV from the other hippocampal subregions from the curve of the CBV generated from the DG. This normalisation is based on the assumption that other factors and parameters, such as cardiac output and synaptic activity, that affect regionally the CBV in the brain, particularly the hippocampus, would have similar effects in the DG and other subregions in the hippocampus [53].

In total, a paradigm has been devised and developed to study adult neurogenesis and its modulation in real time and in live subjects by MRI. This paradigm is based on the tight spatial and temporal coupling between neurogenesis and angiogenesis. It is very useful for studying adult neurogenesis and its modulation in human subjects, as there is basically no alternative at this time, except post-mortem studies. It allows safe, live and real-time studies of adult neurogenesis in animals and humans, an extra benefit of the strategy developed. It may therefore become a very powerful technique to identify and characterise conditions, drugs and molecules that stimulate neurogenesis in the adult hippocampus of humans, in an attempt to compensate or reverse neuronal loss, alleviate symptoms associated with learning and memory deficits, AD and depression, and treat these diseases and disorders [54]. However, there are two main limitations and pitfalls in the application of this paradigm. First, it is indicative of neurogenesis and its modulation. Second, it is based on the assumption that factors and parameters other than adult neurogenesis that affect regionally the CBV in the hippocampus would have similar effects in the DG and other subregions in the hippocampus. These limitations and pitfalls must be assessed carefully and addressed when analysing and discussing the data from these studies. The application claims that the increase of the CBV in the DG, relative to other subregions of the hippocampus, such as the CA1 region, is correlated with neurogenesis in the DG. It claims the use of the imaging of the DG of patients by MRI to monitor neurogenesis and to identify and validate treatments involving stimulation of neurogenesis in the hippocampus.

Declaration of Interest Statement

The author declares no financial and competing interests with the subject matter or materials discussed in the manuscript.

Acknowledgments

Reproduced with permission from Informa Pharmaceutical Science: Taupin P. Magnetic resonance imaging for monitoring neurogenesis in the adult hippocampus. Expert Opinion on Medical Diagnostics 3(2):211-6. Copyright 2009, Informa UK Ltd.

Patent

Title: Imaging correlates of neurogenesis with MRI
Assignee: The Trustee of Columbia University in the City of New York, US
Inventors: Small SA.
Priority data: 14 November 2005
Filing date: 14 November 2006
Publication date: 21 February 2008
Publication no.: WO2008020864

References

Papers of special note have been highlighted as either of interest (•) or of considerable interest (••) to readers.

[1] Taupin P. Neurogenesis in the adult central nervous system. *C. R. Biol.* 2006; 329: 465-75

[2] Duan X, Kang E, Liu CY, et al. Development of neural stem cell in the adult brain. *Curr. Opin. Neurobiol.* 2008; 18: 108-15

[3] Eriksson PS, Perfilieva E, Bjork-Eriksson T, et al. Neurogenesis in the adult human hippocampus. *Nat. Med.* 1998; 4: 1313-7
• First evidence of neurogenesis in the brain of adult humans. Neurogenesis was investigated by mean of BrdU-labelling, immunohistofluorescence and confocal microscopy in the dentate gyrus. The tissue samples originated from cancer patients who were administered BrdU as part of their treatment.

[4] Curtis MA, Kam M, Nannmark U, et al. Human neuroblasts migrate to the olfactory bulb via a lateral ventricular extension. *Science* 2007; 315: 1243-9
• Evidence that neurogenesis occurs in the human subventricular zone. The tissue samples originated from cancer patients who were administered BrdU as part of their treatment.

[5] Stanfield BB, Trice JE. Evidence that granule cells generated in the dentate gyrus of adult rats extend axonal projections. *Exp. Brain Res.* 1988; 72: 399-406

[6] Cameron HA, Woolley CS, McEwen BS, et al. Differentiation of newly born neurons and glia in the dentate gyrus of the adult rat. *Neurosci.* 1993; 56: 337-44

[7] Toni N, Laplagne DA, Zhao C, et al. Neurons born in the adult dentate gyrus form functional synapses with target cells. *Nat. Neurosci.* 2008; 11: 901-7

[8] Taupin P. Adult neurogenesis in the mammalian central nervous system: functionality and potential clinical interest. *Med. Sci. Monit.* 2005; 11: RA247-52

[9] Kuhn HG, Dickinson-Anson H, Gage FH. Neurogenesis in the dentate gyrus of the adult rat: age-related decrease of neuronal progenitor proliferation. *J. Neurosci.* 1996; 16: 2027-33

[10] Kempermann G, Kuhn HG, Gage FH. More hippocampal neurons in adult mice living in an enriched environment. *Nature* 1997; 386: 493-5

[11] Gould E, Tanapat P, McEwen BS, et al. Proliferation of granule cell precursors in the dentate gyrus of adult monkeys is diminished by stress. *Proc. Natl. Acad. Sci. USA* 1998; 95: 3168-71

[12] Brezun JM, Daszuta A. Depletion in serotonin decreases neurogenesis in the dentate gyrus and the subventricular zone of adult rats. *Neurosci.* 1999; 89: 999-1002

[13] Gould E, Beylin A, Tanapat P, et al. Learning enhances adult neurogenesis in the hippocampal formation. *Nat. Neurosci.* 1999; 2: 260-5

[14] van Praag H, Kempermann G, Gage FH. Running increases cell proliferation and neurogenesis in the adult mouse dentate gyrus. *Nat. Neurosci.* 1999; 2: 266-70

[15] Pham K, Nacher J, Hof PR, et al. Repeated restraint stress suppresses neurogenesis and induces biphasic PSA-NCAM expression in the adult rat dentate gyrus. *Eur. J. Neurosci.* 2003; 17: 879-86

[16] Shors TJ, Miesegaes G, Beylin A, et al. Neurogenesis in the adult is involved in the formation of trace memories. *Nature* 2001; 410: 372-6. Erratum in: *Nature* (2001) 414:938

[17] Jin K, Peel AL, Mao XO, et al. Increased hippocampal neurogenesis in Alzheimer's disease. *Proc. Natl. Acad. Sci. USA* 2004; 101: 343-7

• The expression of markers of immature neuronal cells, such as doublecortin and polysialylated nerve cell adhesion molecule, is increased in hippocampal regions, particularly the dentate gyrus, in the brain of patients with AD.

[18] Taupin P. The therapeutic potential of adult neural stem cells. *Curr. Opin. Mol. Ther.* 2006; 8: 225-31

[19] Malberg JE, Eisch AJ, Nestler EJ, et al. Chronic antidepressant treatment increases neurogenesis in adult rat hippocampus. *J. Neurosci.* 2000; 20: 9104-10

[20] Hsieh J, Nakashima K, Kuwabara T, et al. Histone deacetylase inhibition-mediated neuronal differentiation of multipotent adult neural progenitor cells. *Proc. Natl. Acad. Sci. USA* 2004; 101: 16659-64

[21] Jin K, Xie L, Mao XO, et al. Alzheimer's disease drugs promote neurogenesis. *Brain Res.* 2006; 1085: 183-8

[22] Santarelli L, Saxe M, Gross C, et al. Requirement of hippocampal neurogenesis for the behavioral effects of antidepressants. *Science* 2003; 301: 805-9

•• X-ray irradiation of the hippocampal region inhibits neurogenesis and the behavioural activity of antidepressants in adult mice.

[23] Taupin P. Adult neurogenesis pharmacology in neurological diseases and disorders. *Exp. Rev. Neurother.* 2008; 8: 311-20

[24] Taupin P. Adult neurogenesis and drug therapy. Cent Nerv Syst Agents Med Chem 2008; 8: 198-202

[25] Taupin P. Adult neurogenesis and pharmacology in Alzheimer's disease and depression. *J. Neurodegeneration Regen.* 2008; 1: 51-5

[26] Burns A, Byrne EJ, Maurer K. Alzheimer's disease. *Lancet* 2002; 360: 163-5

[27] Taupin P. Adult neurogenesis, neural stem cells and Alzheimer's disease: developments, limitations, problems and promises. *Curr. Alzheimer Res.* 2009 . In press.

[28] Campbell S, Marriott M, Nahmias C, et al. Lower hippocampal volume in patients suffering from depression: a meta-analysis. *Am. J. Psychiatry* 2004; 161: 598-607

[29] Bielau H, Trübner K, Krell D, et al. Volume deficits of subcortical nuclei in mood disorders. A postmortem study. *Eur. Arch Psychiatry Clin. Neurosci.* 2005; 255: 401-12

[30] Taupin P. Neurogenesis and the Effects of Antidepressants. *Drug Target Insights* 2006; 1: 13-7

[31] Miller MW, Nowakowski RS. Use of bromodeoxyuridine-immunohistochemistry to examine the proliferation, migration and time of origin of cells in the central nervous system. *Brain Res.* 1988; 457: 44-52

[32] del Rio JA, Soriano E. Immunocytochemical detection of 5′-bromodeoxyuridine incorporation in the central nervous system of the mouse. *Brain Res. Dev. Brain Res.* 1989; 49: 311-7

[33] Nowakowski RS, Hayes NL. New neurons: extraordinary evidence or extraordinary conclusion? *Science* 2000; 288: 771

[34] Gould E, Gross CG. Neurogenesis in adult mammals: some progress and problems. *J. Neurosci.* 2002; 22: 619-23

[35] Taupin P. BrdU Immunohistochemistry for Studying Adult Neurogenesis: paradigms, pitfalls, limitations, and validation. *Brain Res. Rev.* 2007; 53: 198-214

[36] Taupin P. Protocols for studying adult neurogenesis: insights and recent developments. *Regen. Med.* 2007; 2: 51-62

[37] Bhardwaj RD, Curtis MA, Spalding KL, et al. Neocortical neurogenesis in humans is restricted to development. *Proc. Natl. Acad. Sci. USA* 2006; 103: 12564-8

[38] González RG, Fischman AJ, Guimaraes AR, et al. Functional MR in the evaluation of dementia: correlation of abnormal dynamic cerebral blood volume measurements with changes in cerebral metabolism on positron emission tomography with fl udeoxyglucose F 18. *AJNR Am. J. Neuroradiol.*1995; 16: 1763-70

[39] Small SA, Wu EX, Bartsch D, et al. Imaging physiologic dysfunction of individual hippocampal subregions in humans and genetically modifi ed mice. *Neuron* 2000; 28: 653-64

•• MRI has sufficient spatial resolution to visualise regions and subregions of the brain, such as the hippocampus and dentate gyrus, in humans and rodents.

[40] Small SA, Tsai WY, DeLaPaz R, et al. Imaging hippocampal function across the human life span: is memory decline normal or not? *Ann. Neurol.* 2002; 51: 290-5

[41] Guzman R, Uchida N, Bliss TM, et al. Long-term monitoring of transplanted human neural stem cells in developmental and pathological contexts with MRI. Proc. *Natl. Acad. Sci. USA* 2007; 104: 10211-6

[42] Shapiro EM, Sharer K, Skrtic S, et al. In vivo detection of single cells by MRI. *Magn. Reson. Med.* 2006; 55: 242-9

[43] Shapiro EM, Gonzalez-Perez O, Manuel García-Verdugo, et al. Magnetic resonance imaging of the migration of neuronal precursors generated in the adult rodent brain. *Neuroimage* 2006; 32: 1150-7

• MRI technique developed and applied to study newly generated neuronal cells in the adult brain of rodents.

[44] Pereira AC, Huddleston DE, Brickman AM, et al. An in vivo correlate of exercise-induced neurogenesis in the adult dentate gyrus. *Proc. Natl. Acad. Sci. USA* 2007; 104: 5638-43

[45] Taupin P, Ray J, Fischer WH, et al. FGF-2-responsive neural stem cell proliferation requires CCg, a novel autocrine/paracrine cofactor. *Neuron* 2000; 28: 385-97

[46] Wagner PD. Determinants of maximal oxygen transport and utilization. *Annu. Rev. Physiol.* 1996; 58: 21-50

[47] Palmer TD, Willhoite AR, Gage FH. Vascular niche for adult hippocampal neurogenesis. *J. Comp. Neurol.* 2000; 425: 479-94

[48] Louissaint A Jr, Rao S, Leventhal C, et al. Coordinated interaction of neurogenesis and angiogenesis in the adult songbird brain. *Neuron* 2002; 34: 945-60

[49] Taupin P. Adult neural stem cells, neurogenic niches and cellular therapy. *Stem Cell Rev.* 2006; 2: 213-20

[50] Ninkovic J, Götz M. Signaling in adult neurogenesis: from stem cell niche to neuronal networks. *Curr. Opin. Neurobiol.* 2007; 17: 338-44

[51] Jordan JD, Ma DK, Ming GL, et al. Cellular niches for endogenous neural stem cells in the adult brain. *CNS Neurol. Disord. Drug Targets* 2007; 6: 336-41

[52] Dunn JF, Roche MA, Springett R, et al. Monitoring angiogenesis in brain using steady-state quantification of DeltaR2 with MION infusion. *Magn. Reson Med.* 2004; 51: 55-61

[53] Delp MD, Armstrong RB, Godfrey DA, et al. Exercise increases blood fl ow to locomotor, vestibular, cardiorespiratory and visual regions of the brain in miniature swine. *J. Physiol.* 2001; 533: 849-59

[54] Taupin P. Therapeutic neuronal stem cells: patents at the forefront. *Expert Opin. Ther. Patents* 2008; 18: 1107-10

Chapter XII

HuCNS-SC StemCells

Abstract

HuCNS-SC, a proprietary human neural stem cells product, is being developed as a cellular therapy for the potential treatment of Batten disease, one of a group of disorders known as neural ceroid lipofuscinoses (NCL). Developer StemCells is also investigating the therapy for spinal cord injury and other central nervous system disorders, such as demyelinating disease, stroke and Alzheimer's disease. A phase I trial of HuCNS-SC for infantile and late-infantile NCL has been initiated, following the March 2006 US Food and Drug Administration approval of StemCells' investigational new drug application.

Summary

* Originator NeuroSpheres Ltd
* Licensee StemCells Inc
* Status Phase I Clinical
* Actions Nootropic agent
* Indications Central nervous system disease, Lysosome storage disease, Neurodegenerative disease, Spinal cord injury
* Technology Stem-cell therapy

Introduction

Cellular therapy can be summarized as the replacement of dysfunctional or degenerated tissues by new ones, and as such, carries a lot of hope for the treatment of a broad range of diseases and injuries, particularly of the central nervous system (CNS). Cell types from various sources are currently being evaluated for cellular therapy, including embryonic stem cells (ESC). ESC are derived from the inner cell mass of blastocytes and, because they have the potential to generate virtually all the cell types found within the body, major efforts are being devoted to bringing ESC to the clinic. Although ESC have been popularized as a potential therapeutic breakthrough, there are limitations to the use in cellular therapy, the major drawback being the risk of undifferentiated ESC developing into tumor tissue upon grafting [647207].

Neural stem cells (NSC) offer an alternative to ESC. NSC are the self-renewing, multipotent cells that generate the main phenotypes of the nervous system: neurons, astrocytes and oligodendrocytes [437208]. NSC are present in both fetal and adult mammalian tissues, from which they can be isolated and cultured *in vitro*, providing a source of tissue for cellular therapy [647230]. NSC therapy is currently being proposed for the treatment of a broad range of CNS diseases and injuries, in particular neurodegenerative diseases, spinal cord injuries, stroke and genetic disorders that damage the brain [655176]. It is possible that in the future, therapies based on NSC could aid the reconstruction of damaged brain tissues; however, a great deal of research must be done before medicine advances to this point.

Batten's disease (BD) is one of a group of disorders known as neural ceroid lipofuscinoses (NCL). NCL are inherited autosomal recessive neurodegenerative disorders of the nervous system that usually occur in childhood. They are caused by an abnormal build-up of substances called lipofuscins in the nerve cells throughout the brain, which leads to a progressive deterioration of brain function [647967]. BD affects the nerve cells in the brain and eyes, as well as other parts of the body, causing a progressive loss of vision, decline in physical and mental capabilities and seizures. Early symptoms usually appear between the ages of five and ten years, and the disease is often fatal by the late teens or twenties. Lipofuscinoses are normally broken down and removed from the body, but in BD, various genes responsible for this process are faulty. An enzyme called palmitoyl-protein thioesterase 1 (PPT1) is insufficiently active in the infantile form of BD, because of a mutation in the gene encoding ceroid lipofuscinosis neural 1. PPT1 is a lysosomal protein that removes fatty acyl side chains from cysteine residues of proteins in lysosomes. Deficiency of PPT1 causes an abnormal build-up of substances in the nerve cells, and subsequently leads to a decline in nervous system function [647967]. PPT1 knockout mice have been generated as a model of BD. PPT1-/- mice present neuronal loss in the hippocampus and cerebellum, as well as an accumulation of auto-fluorescent storage material within these areas [647969].

BD and other forms of NCL are relatively rare, occurring in an estimated two to four of every 100,000 live births in the US [www.ninds.nih.gov]. There is currently no specific treatment for BD and current therapy simply alleviates the symptoms of the disease. Anticonvulsant drugs alleviate the associated seizures, and occupational therapy helps individuals compensate for the loss of vision, physical and mental abilities. Because BD involves the deterioration of neuronal cell tissue, it is a candidate for cellular therapy.

StemCells Inc is developing a proprietary NSC product for cellular therapy, under license from NeuroSpheres Ltd, comprising well-characterized, normal human CNS stem cells (HuCNS-SC) from brain tissue. HuCNS-SC is currently under investigation for the potential treatment of neurodegenerative disorders, particularly BD [180367], [540074]. Preclinical studies have been performed in various animal models of CNS diseases and injuries. Data from these studies has supported the therapeutic potential of HuCNS-SC, and the therapy has recently been approved for a phase I clinical trial for the treatment of BD [629732].

Synthesis and SAR

In 1992, Reynolds and Weiss invented a reproducible method for growing rodent CNS stem and progenitor cells in culture as clusters of cells called neurospheres [200019••]; in 1997, the procedure was adapted for human CNS stem and progenitor cells [437374]. HuCNS-SC were derived from human fetal (16 to 20 weeks) brain tissue. After enzymatic dissociation of the tissue into single-cell suspension, cells were isolated by fluorescence-activated cell sorting using novel combinations of antibodies to cell-surface markers [437290]. The isolated cells were cultured in neurosphere conditions, in defined medium, with a cocktail of trophic factors containing basic fibroblast growth factor, epidermal growth factor, lymphocyte inhibitory factor, neural survival factor-1 and N-acetylcysteine. *In vitro* studies conducted on these cells demonstrated that CD133+ cells were capable of neurosphere initiation, self-renewal and multi-potentiality, thus making them the ideal HuCNS-SC [437290]. CD133 is a major NSC marker that is defined by its five transmembrane domains (a unique structure among known cell-surface markers), and antibodies to CD133 (such as the monoclonal antibody AC133 [647217]) have previously been used to isolate human hematopoietic cells and ESC [647217].

Sorting procedures further characterized the HuCNS-SC population as CD133+, 5E12+, CD34-, CD45- and CD24-/lo (CD133+ sorted cells). After 8 weeks, 5 to 10% of single sorted CD133+ cells developed into neurospheres, whereas the sorted cell population not expressing CD133 (CD133-, 5E12+, CD34-, CD45-and CD24-/lo), representing approximately 95% of total fetal brain tissue, failed to differentiate into neurospheres. CD133+ sorted cells were grown and expanded in culture through cell dissociation and splitting techniques. After five passages, the number of CD133+ cells increased by at least 1000-fold, and cultured cells retained their ability to re-initiate neurosphere formation. When plated in differentiating medium containing two growth factors (brain-derived neurotrophic factor and glial-derived neurotrophic factor), clonally derived neurospheres from CD133+ cells differentiated into neurons and astrocytes. Together, these data demonstrated that CD133+ sorted cells represent a population of cells with stem-cell properties, and that highly enriched NSC populations could be isolated with antibodies against CD133 [437290].

Preclinical Development

Initial studies by CytoTherapeutics Inc (later StemCells Inc), which led to the development of HuCNS-SC, focused on the delivery of protein neurotrophic factors into the neural system. In particular, studies focused on the transfer of human nerve growth factor (NGF) into neurodegenerative animal models [199982], [199983], [225755], [243560], [243561], [243562]. Baby hamster kidney cells were genetically modified to secrete high levels of human NGF before undergoing polymer encapsulation prior to implantation. Although the therapy showed potential for the treatment of neurological disorders, the emergence of stem cells offered far broader therapeutic scope.

Animal NSC

The potential of stem-cell therapy was examined *in vitro* using murine fetal NSC. The stem cells were epidermal growth factor (EGF)-responsive, that is, were continuously propagated *in vitro* in the presence of EGF. In the absence of EGF and in the presence of 1% fetal calf serum, the NSC differentiated to form astrocytes, oligodendrocytes and neurons. After 5 to 10 days, the promoter elements glial fibrillary acidic protein (GFAP) and myelin basic protein (MBP) were expressed on astrocytes and oligodendrocytes, respectively [200021], [243563]. GFAP and MBP were used to direct the expression of the Escherichia coli reporter gene ßgalactosidase in a transgenic marking system in mice. Studies monitoring expression of ß-galactosidase found the gene was highly stable, as it appeared to be expressed in virtually 100% of the appropriate cells [200021].

One of the first animal studies of NSC showed that rodent EGF-responsive NSC could differentiate into oligodendroglia. Myelin-deficient rats, killed 13 to 14 days post-treatment, developed patches of myelin around the dorsal columns, indicating a possible valuable source of myelinating cells [200023]. Further studies with murine NSC looked at the potential of these cells to alleviate excitotoxic striatal lesions in rats. Murine fetal stem cells containing the human GFAP promoter element were transplanted into the striatum. Animals receiving control cells demonstrated comprehensive lesions of the striatum compared with animals receiving the GFAP stem cells. The GFAP stem cells promoted the expression of human nerve growth factor (NGF), which helped to protect against striatal lesions [243558].

HuCNS-SC

An initial study of HuCNS-SC assessed the potential of the cells to self-renew in non-obese diabetic/severe combined immunodeficient (NOD/SCID) mice transplanted 1 to 12 months prior to analysis. HuCNS-SC from transplanted murine brain were re-isolated by immunomagnetic selection for human specific neural cell adhesion molecules and thymus cell antigen 1 to assess the self-renewal of the transplanted cells. Some of the cells expressed the human specific antigens AC133 or hCD24 [468477]. After transplantation, most cells were AC133-/hCD24hi, but some AC133+/hCD24-/lo antigens were observed; the sorted AC133+/hCD24-/lo human cells from one-year post-transplant re-initiated neurosphere cultures, indicating that HuCNS-SC continue to generate more AC133+ cells as well as give rise to AC133-/hCD24hi progenitors for up to one year following transplantation [468477].

An objective comparison of engraftment data between different HuCNS-SC-derived neurosphere lines showed that, *ex vivo*, expanded HuCNS-SC engraft robustly in the NOD/SCID mouse brain and retain a multilineage differentiation capacity. Newborn mice were injected with three neurosphere cell lines (1.55 HuCNS-SC per hemisphere). Stereological analysis estimated that 54 to 74 transplanted cells were distributed to the cortex of one hemisphere. Further analysis using confocal microscopy showed that HuCNS-SC gave rise to neurons, astrocytes and oligodendrocytes in a site-specific manner [636344]. This study followed on from earlier studies that had demonstrated the ability of HuCNS-SC to engraft, proliferate, migrate and differentiate into neurons, astrocytes and oligodendrocytes for at least 7 months post-transplantation into the brains of NOD/SCID mice [437290], [647222].

The survival and expression of mouse NSC and HuCNS-SC (10 to 50 × 106 cells per ml) were compared in non-lesioned, Parkinsonian or Huntingtonian rat hosts. Analyses were conducted over 1 week to 12 months and demonstrated cell survival of 2 to 10% (mouse) and 10 to 35% (human). The transplanted cells differentiated into both neurons and astrocytes and migrated away from the injection tract in a radial fashion, extending 0.5 to 1.5 mm mediolaterally and rostrocaudally [243557].

HuCNS-SC were transplanted into the ischemic cortex of rats 7 days after distal middle cerebral artery occlusion (MCAO), because the lesion sizes were relatively stable by this time, allowing for specific targeting of the peri-infarct area. Rats were administered three 1.0-μl deposits of suspended cells (1 × 105 cells per μl) along the anterior-posterior axis into the cortex via an infusion pump. The brains of the rats were analyzed 4 weeks later. Positive survival of cells was observed only if the cells were transplanted into non-ischemic tissue. There was a negative correlation of lesion size with cell survival, suggesting that even 7 days after MCAO the environment is unfavorable to the transplanted cells. It was suggested that inflammatory cytokines might prevent the transplanted cells integrating and surviving in the lesioned tissue, and that antiinflammatory treatments might be beneficial in assisting cellular therapy. The cells also demonstrated targeted migration, with human cells mainly differentiating to the neuronal phenotype and migrating long distances (~ 1.2 mm compared with 0.2 mm for naïve rats), predominantly toward the lesion [647228], [654203].

Remyelination by HuCNS-SC was observed in both spinal cord injury NOD/SCID and myelin-deficient shiverer mice (defective in myelin production as a result of a mutation in MBP) [623960••], [633798], [636193], [654205]. HuCNS-SC injected into NOD/SCID spinal cord injured mice 9 days after contusion, survived, migrated and expressed differentiation markers for neurons and oligodendrocytes. Electron microscopy analysis gave evidence consistent with synapse formation between HuCNS-SC and mouse host neurons, although no details of presynaptic output or electrophysiological evidence were presented. However, HuCNS-SC did not contribute to glial scar formation because of glial cell proliferation, suggesting that few engrafted human cells differentiated into astrocytes in this model. The mice received four injections bilaterally of a 250nl cell suspension, at 75,000 cells per μl. By 17 weeks posttreatment, cells had migrated away from the lesion epicenter in sagittal sections, with some transplanted cells found > 1 cm from the lesion epicenter. The expression of ß-tubulin III (expressed by neurons early in their development) was observed in approximately 26% of the implanted cells [623960••], [654204], [654205].

In shiverer mice, data indicated the presence of proliferating cells because of the high expression of the proliferation marker Ki-67+ at early time points. Only a limited number of HuCNS-SC expressed the NSC marker CD133; however, expression of nestin (an intermediate filament protein expressed in neuroepithelial stem cells during nervous system development) was prevalent. Instead of differentiating into neurons, histological assessment indicated that HuCNS-SC largely differentiated into oligodendrocytes through a continuous process of differentiation lasting approximately 45 days until full maturation and MBP production. By the end of the study (60 days), all grafted white matter tracts in the cerebellum, fimbria and corpus callosum demonstrated dense MBP staining. However, transplanted cells within the cortex and the hippocampus, and cells remaining in the injection core, failed to express MBP by day 60. Immunoelectromagnetic analysis, using the human specific monoclonal antibody SC121 showed that the oligodendrocytes derived from HuCNS-SC enrobed murine axons with 10 to 16 layers of myelin [633798], [636193].

HuCNS-SC were used as a cellular delivery vehicle for enzyme replacement in a murine model (PPT1 knockout mice) of the infantile form of BD. HuCNS-SC were cocultured with fibroblasts taken from BD individuals and used to produce and secrete the PPT1 enzyme. After cells were transplanted into the ventricle of mice, preliminary data demonstrated a reduced build-up of pathogenic toxic waste material and a larger number of surviving neurons in transplanted transgenic mice compared with control [485519]. Follow-up data from the study at 18 weeks post-transplant demonstrated that the PPT1 enzyme was replaced by HuCNS-SC. HuCNS-SC had migrated in a site-specific manner throughout the mouse brain and caused a reduction (in both number and intensity) in lysosomal storage material associated with disease pathology, compared with non-transplanted, age-matched controls [570810], [654206].

Further follow-up data showed the cells to engraft and secrete enzymes for up to 6 months. The brains of treated mice expressed signs of decreased lysosomal storage material in a number of cerebral regions ($p < 0.0001$). HuCNS-SC demonstrated dose-dependent neuroprotection (33 and 54% of neurons being scored as normal following low- and high-dose transplantation, respectively, compared with 8% in control). Enzyme assays revealed that the high dose of HuCNS-SC caused a significant increase in enzyme levels 160 to 190 days after transplantation, to a level that was thought to be above the assumed threshold level for symptomatic disease in humans. These data indicate that grafted HuCNS-SC survive robustly in diseased and injured brains, migrate to the site of degeneration where they differentiate into neural lineages, and have a beneficial effect on functional recovery [633798], [654204].

Metabolism and Pharmacokinetics

No data are currently available.

Toxicity

There were no signs of tumor formation at one year post-injection of HuCNS-SC (105 or 106 cells) in NOD/SCID mice [437290]. No further data are currently available on the toxicity of HuCNS-SC in animal models.

Clinical Development

In January 2005, StemCells filed an investigational new drug (IND) application for a phase I clinical trial in infantile and late-infantile BD [578153], but in February 2005, the

Food and Drug Administration (FDA) placed the IND on hold while it raised a number of queries with StemCells regarding the trial [582789]. This application was the first time that the FDA had been asked to review a proposed clinical trial involving the use of purified human NSC as a potential therapeutic agent. Later in the year, StemCells submitted amendments to the IND, including plans for patient committal to a four-year follow-up study, and the trial was approved by the FDA in October 2005 [629732].

The open-label, phase I/II trial is planned to enroll patients in the advanced stage of infantile BD, or late-infantile disease, and will evaluate two doses of HuCNS-SC, which will be injected directly into the brain. The primary objective of the trial is to measure the safety (adverse effects, brain magnetic resonance imaging) of HuCNS-SC, and to evaluate the ability of HuCNS-SC to affect disease progression (cognition, communication, behavior, motor function) and quality of life (disability, behavior, communication, general health, seizures). Patients are expected to undergo continuous immunosuppression [578153], [633798].

Side Effects and Contraindications

A major potential flaw in the future development and therapeutic use of HuCNS-SC is the risk of immune rejection because the stem cells are derived from fetal donors. This risk has not been addressed in the studies reported to date relating to the use of HuCNS-SC. Allografts require immunosuppressive treatments, such as cyclosporine, to prevent host rejection. These treatments are not very well tolerated by patients, producing side effects such as renal dysfunction, tremor and hypertension. This may well affect the potential benefits of HuCNS-SC therapy.

Patent Summary

The culture system used by StemCells to develop HuCNS-SC therapy was first disclosed in 1993 in the patent application WO-09301275, assigned to S Weiss and BA Reynolds, the founders of NeuroSpheres. Several additional patent applications have been filed by NeuroSpheres, including WO-09409119, claiming a method to remyelinate neurons using NSC; WO-09513364, disclosing methods for inducing proliferation and/or differentiation of neural progenitor cells both *in vivo* and *in vitro*; WO-09410292, disclosing methods for proliferating and differentiating NSC into astrocytes, oligodendrocytes or neurons; US-05750376, covering the *in vitro* growth and proliferation of genetically modified NSC; and WO-09615226, covering the regulation of NSC proliferation.

CytoTherapeutics filed two patent applications covering the use of NSC. WO-09911758 disclosed the isolation, characterization, proliferation, differentiation and transplantation of mammalian NSC, and WO-00050572 covers a method for the *in vitro* proliferation of NSC cultures using a growth factor and a collagenase enzyme said to increase cell viability and the number of proliferative cells with time.

StemCells have filed WO-2004020597 disclosing methods for identifying, isolating and enriching CNS stem and progenitor cell populations. The patent covers methods that utilize reagents, such as immunoglobulins, which bind cell-surface markers, such as CD49f, CD133 and CD15. StemCells claims to hold 42 patents in the US and 108 patents worldwide ('in 14 foreign equivalent cases'), with a further 80 applications pending as of October 2005 [www.stemcellsinc.com].

Current Opinion

The preclinical data reported to date show that grafted HuCNS-SC can survive in damaged brain tissue and migrate to the specific site of degeneration, where they differentiate into neural lineages, and have a beneficial effect on functional recovery. The current evidence emphasizes the favorable therapeutic potential of HuCNS-SC therapy. However, additional *in vivo* and preliminary clinical studies are necessary to obtain a better understanding of the mechanisms of regeneration and recovery, and to further evaluate the potential therapeutic benefit of HuCNS-SC.

Although the data shows the potential and promise of HuCNS-SC therapy for the treatment of CNS diseases and injuries, there is one major concern: the risk of tissue rejection of the fetally derived cells in allograft transplantation. Preventing tissue rejection would require either genetically matching the donor with recipient, or administering life-long immunosuppressive treatments, such as cyclosporine. On the one hand, optimal donor-recipient matching would require establishing HuCNS-SC banks, representing both a technical challenge and an ethical hurdle because of the stringent regulations governing the use of fetal tissues for therapeutic research. On the other hand, the toxicity of immunosuppressive drugs is well established, thereby limiting the use of HuCNS-SC for therapy.

There is also one unknown parameter regarding the use of HuCNS-SC for therapy that has not been addressed in preclinical studies so far. Most of the studies were performed in immunosuppressed rats or NOD/SCID mice. Therefore, the activity of cytokines and chemokines of the immune system and their potential risks on HuCNS-SC therapy have not been fully investigated. Stem cells, including neurospheres, respond to the expression and activities of cytokines and chemokines. These activities may have adverse effects on the survival, migration and differentiation of the HuCNS-SC and thus affect their potential therapeutic use.

Nonetheless, a range of studies has confirmed the potential of NSC for cellular therapy in the CNS (as reviewed in [243564], [284931]). The mechanisms underlying the recovery of neural progenitor and stem cells after grafting are yet to be fully characterized. The synthesis and release of neuroprotective substances by the grafted neural progenitor and stem cells have been proposed as a likely mechanism of the functional recovery [647946], [647954], but the recent of remyelination by HuCNS-SC in both spinal cord injury NOD/SCID mice and myelin-deficient shiverer mice demonstrated that grafted neural progenitor and stem cells can also contribute to the recovery by their integration in the CNS network [623960••]. Both of these properties may be beneficial for the treatment of the diseased and injured brain, and particularly for neurodegenerative diseases such as BD. Results also demonstrate how neural progenitor and stem cells migrate toward the site of injury and degeneration when transplanted in the CNS (administered either by systemic injection, or through the cerebrospinal fluid), making them particularly suitable for the treatment of neurodegenerative diseases where the degeneration is widespread, such as in BD, Alzheimer's disease and Huntington's disease [654686], [654690]. More specifically, the recovery of myelination within animal models suggests a possible treatment for multiple sclerosis or periventricular leukomalacia (PVL), a kind of cerebral palsy caused by errors in myelination [654691].

Neural stem and progenitor cells can also be genetically modified [437418], [US-05750376], extending their potential use for the treatment of neurological diseases caused by genetic deficiencies, but also to promote neuronal survival in neurodegenerative diseases. Genetically modified neural progenitor and stem cells have been proposed for the treatment of type A Niemann Pick disease, a lysosomal storage disorder in which deficiency of acid sphingomyelinase leads to the intracellular accumulation of sphingomyelin and cholesterol in lysosomes [647908]. Genetically engineered neural progenitor and stem cells expressing acid sphingomyelinase have been reported to reverse lysosomal storage pathology in animal models of Niemann Pick disease [647919], confirming the potential of NSC to serve as a gene transfer vehicle for the treatment of CNS pathology, and particularly for lysosomal storage diseases. Such a strategy may be applicable to BD, by genetically engineering neural progenitor and stem cells to express PPT1 [570810]. These observations demonstrate that there may be various alternative treatments for the diseased and injured brain using NSC therapy.

With the recent confirmation that neurogenesis occurs in the adult brain and that NSC reside in the adult CNS, new opportunities for cellular therapy in the CNS are being considered [437208], [647230]. Cellular therapeutic intervention may involve the stimulation of endogenous progenitor cells, or the transplantation of neural progenitor and stem cells derived from the adult brain. Neural progenitor and stem cells may also be isolated from biopsies and post-mortem tissues, providing multiple sources of tissues for therapy [437394•], [647233]. Such strategies carry a lot of hope for the treatment of CNS diseases and injuries, and may provide an alternative to the use of fetally derived neural progenitor and stem cells, thus lacking the ethical and political constraints associated with fetally derived tissues. There are, however, several questions that need to be addressed before NSC technology can be brought to therapy: how will the selective differentiation of NSC toward the desired phenotype(s), particularly in an environment that may not be favorable, be controlled, directed and optimized? Will the new neuronal cells establish the right CNS connections? What are the chances that they will establish connections with the wrong target cells? A process that maintains the developmental potential of NSC will have to be devised and validated to produce the large quantities of NSC required for therapy. NSC therapy still needs further characterization, particularly with regard to the relationship of these cells with tumor cells, before any broader evaluations of its therapeutic use for treating neurodegenerative diseases can be started.

Licensing

Ciba-Geigy AG

By 1995, Ciba-Geigy (now Novartis) and NeuroSpheres had entered into a collaborative research agreement for the treatment of CNS disorders with stem cells; however, no development has been reported on the project by Ciba-Geigy since then [200025].

StemCells Inc

In March 1994, CytoTherapeutics Inc (now StemCells) entered into a research and license agreement with NeuroSpheres for the development of therapeutic products involving encapsulated and non-encapsulated neural stem cells. NeuroSpheres granted CytoTherapeutics an exclusive, worldwide, royalty-bearing license to develop and sell products involving the transplantation of neural stem cells developed by NeuroSpheres [180367]. This agreement was further clarified in April 1997, whereby CytoTherapeutics obtained an exclusive patent license from NeuroSpheres in the field of neural stem-cell transplantation. NeuroSpheres had an option to buy a non-exclusive license which was not exercised. An additional agreement was signed with NeuroSpheres in October 2000, for an exclusive license to non-transplant uses of the cells. Upfront payments to NeuroSpheres of 65,000 shares of CytoTherapeutics common stock were made in October 2000, and US $50,000 in January 2001. In October 2000, CytoTherapeutics reimbursed NeuroSpheres for patent costs amounting to $341,000. Annual payments of $50,000 a year were to be made to NeuroSpheres beginning in 2004 [560410].

Development History

In September 2003, StemCells was awarded a 1-year, US $0.34 million SBIR grant from the National Institute of Neurological Disease and Stroke to further work in the treatment of spinal cord injuries [504337]. One year later, the National Institutes of Health awarded two grants focusing on the use of StemCells' human NSC. The company was awarded a Small Business Technology Transfer grant of $0.47 million for studies in Alzheimer's disease, to be conducted by the McLaughlin Research Institute, and the Reeve-Irvine Center at the University of California Irvine received a multiyear grant of $1.4 million to fund studies on human CNS stem-cell grafts in the treatment of spinal cord injuries [559872]. In August 2005, StemCells received a manufacturing license for its cell processing facility in California, allowing the company to conduct clinical trials of HuCNS-SC [620544].

Developer	Country	Status	Indication	Date	Reference
StemCells Inc	US	Phase I	Lyosome storage disease	09-MAR-06	654808
StemCells Inc	US	Discovery	Central nervous system disease	14-JUL-95	180367
StemCells Inc	US	Discovery	Spinal cord injury	09-JUN-04	540074
NeuroSpheres Ltd	Canada	Discontinued	Neuro-degenerative disease	01-APR-97	560410
Ciba-Geigy AG	Switzerland	No development reported	Central nervous system disease	25-SEP-95	-

Literature classifications

Chemistry		
Study type	Result	Reference
HuCNS-SC isolation and differentiation.	Antibody-sorted cells expressing the stem-cell marker CD133 increased ~ 1000-fold in population after five passages. CD133 cells retained their ability to re-initiate neurosphere formation when plated in differentiating media	437290

Biology				
Study type	Effect studied	Model	Result	Reference
In vivo	Engraftment, proliferation, migration and differentiation	HuCNS-SC (10^5 or 10^6 cells) were transplanted into the brains of NOD/SCID mice	HuCNS-SC expressed potent engraftment, proliferation, migration and neural, differentiation, for at least 7 months post-transplantation	437290
In vivo	Survival, migration and differentiation	Three deposits of 10^5 HuCNS-SC were transplanted into the ischemic cortex of rats 7 days after MCAO	Analysis at 4 weeks showed that the lesion size was negatively correlated to cell survival. The cells migrated up to 1 mm more than control towards the lesion, mainly differentiating to the neuronal phenotype.	647228
Ex vivo	Locomotor recovery and cell migration	Four injections of 3^5 HuCNS-SC were injected into spinal cord injured NOD/SCID mice 9 days after contusion	HuCNS-SC engraftment was associated with locomotor recovery at 16 weeks post-treatment. By 17 weeks post-treatment, cells had migrated away from the lesion epicenter, in sagittal sections, with some transplanted cells found > 1 cm from the lesion epicenter	623960••

Literature classifications (Continued)

Biology				
Study type	Effect studied	Model	Result	Reference
Ex vivo	Enzyme Production	HuCNS-SC co-cultured with fibroblasts from BD individuals were injected into a mouse model(PPT1 knockout mice) of BD	HuCNS-SC engrafted and secrete enzymes for up to6 months. Enzyme assays revealed that the high doses of HuCNS-SC caused a significant increase in enzyme levels160 to 190 days after transplantation. The brains of treated mice demonstrated signs of decreased lysosomal storage material and increased neuroprotection	633798
Ex vivo	MBP production	HuCNS-SC were injected into myelin-deficient shiverer mice	HuCNS-SC largely differentiated into oligodendrocytes through a continuous process of differentiation lasting ~ 45 days until full maturation and MBP production. By the end of the study (60 days), all grafted white matter tracts in the cerebellum, fimbria and corpus callosum demonstrated dense MBP staining. The oligodendrocytes enrobed murine axons with 10 to 16 layers of myelin	636193

Associated patent

Title Novel growth factor-responsive progenitor cells which can be proliferated (*in vitro*).
Assignee Individual
Publication WO-09301275 21-JAN-93
Priority US-1990726812 08-JUL-91
Inventors Weiss S, Reynolds BA.

Acknowledgments

Reproduced, with permission from Thomson Reuters and Taupin P: HuCNS-SC (StemCells). Current Opinion in Molecular Therapeutics (2006) 8(2):156-63. Copyright 2006, Thomson Reuters (Scientific) Ltd.

References

•• of outstanding interest
• of special interest

180367 Annual Report 1994 – CytoTherapeutics Inc. CytoTherapeutics Inc *ANNUAL REPORT* 1994 December 31

199982 Transplantation of polymer-encapsulated cells genetically modified to secrete human nerve growth factor prevents the loss of degenerating cholinergic neurons in rats and nonhuman primates. Emerich DF, Winn SR, Hammang JP, Kordower JH, Frydel BR, Baetge EE *CELL TRANSPLANT* 1994 3 3 212

199983 Implants of polymer-encapsulated human NGF-secreting cells in the nonhuman primate: Rescue and sprouting of degenerating cholinergic basal forebrain neurons. Emerich DF, Winn SR, Harper J, Hammang JP, Baetge EE, Kordower JH J *COMP. NEUROL* 1994 349 1 148164

200019 Generation of neurons and astrocytes from isolated cells of the adult mammalian central nervous system. Reynolds BA, Weiss S *SCIENCE* 1992 255 5052 1707-1710

•• This study reports the first isolation and characterization of neural progenitor and stem cells from the adult brain. Neural progenitor and stem cells were isolated from mouse striatal tissue containing the subventricular zone (a neurogeneic area of the adult brain) and cultured in vitro in the presence of epidermal growth factor.

200025 Does Ciba need a clearer strategic direction? *MARKETLETTER* 1995 September 25

200021 Cell-specific expression of reporter genes in EGF-responsive neural stem cells derived from transgenic mice. Hammang JP, Duncan ID, Messing A *CELL TRANSPLANTATION* 1994 3 3 230

200023 EGF-responsive neural stem cells isolated from rat and mouse brain are capable of differentiating into oligodendrocytes and of forming myelin following transplantation into the myelin deficient rat. Duncan ID, Archer DR, Hammang JP *ABSTR. SOC. NEUROSCI.* 1993 19 1-3 Abs 689

225755 Polymer-encapsulated genetically modified cells continue to secrete human nerve growth factor for over one year in rat ventricles: Behavioral and anatomical consequences. Winn SR, Lindner MD, Lee A, Haggett G, Francis JM, Emerich DF *EXP. NEUROL.* 1996 140 2 126-138

243557 Progeny of long-term mouse and human CNS stem cell cultures survive transplantation and differentiate into neurons and glia. Borg LL, Vescovi AL, Blote K, Kyle AL, Hettiaratchi P, Reynolds BA, Mudrick-Donnon LA *ABSTR. SOC. NEUROSCI.* 1996 22 1-3 47

243558 Transplantation of EGF-responsive neural stem cells derived from GFAP-hNGF transgenic mice attenuates excitotoxic striatal lesions. Carpenter MK, Winkler C, Fricker R, Wong SC, Greco C, Emerich D, Chen EY, Chu Y, Kordower J, Messing A, Bjorklund A, Hammang JP *ABSTR. SOC. NEUROSCI.* 1996 22 1-3 577

243560 Long-term expression of nerve growth factor from polymer encapsulated cells in the rat CNS. Hammang JP, Dean BJ, Lee A, Emerich DF, Winn SR, Bamber B, Palmiter RD, Baetge EE *ABSTR. SOC. NEUROSCI.* 1993 19 1-3 657

243561 The delivery of neurotrophic factors to the nervous system using polymer encapsulated cells. Baetge EE, Winn SR, Lee A, Dean BJ, Bamber B, Palmiter RD, Hammang JP *ABSTR. SOC. NEUROSCI.* 1993 19 1-3 657

243562 Delivery of a putative Parkinson's factor (GDNF) into the rat CNS using a polymer-encapsulated cell line. Baetge EE, Emerich DF, Winn SR, Lee A, Lindner MD, Hammang JP *MOL. BIOL. CELL* 1993 4 Suppl 442A

243563 EGF-responsive neural stem cells derived from MBP-lacZ transgenic mice can express the transgene in differentiating oligodendrocytes. Hammang JP, Wrabetz LG, Kamholz J, Messing A *MOL. BIOL. CELL* 1993 4 Suppl 374A

243564 Neural stem cells for CNS transplantation. Baetge EE *ANN. N.Y. ACAD. SCI.* 1993 695 285-291

284931 Treatment of neurodegenerative diseases with neural cell transplantation. Dinsmore JH *EXPERT OPIN. INVEST. DRUGS* 1998 7 4 527-534

437208 Mammalian neural stem cells. Gage FH *SCIENCE* 2000 287 5457 1433-1438

437290 Direct isolation of human central nervous system stem cells. Uchida N, Buck DW, He D, Reitsma MJ, Masek M, Phan TV, Tsukamoto AS, Gage FH, Weissman IL *PROC. NATL. ACAD. SCI. USA* 2000 97 26 14720-14725

437374 In vitro expansion of a multipotent population of human neural progenitor cells. Carpenter MK, Cui X, Hu Z-Y, Jackson J, Sherman S, Seiger A, Wahlberg LU *EXP. NEUROL* 1999 158 2 265-278

437394 Cell culture. Progenitor cells from human brain after death. Palmer TD, Schwartz PH, Taupin P, Kaspar B, Stein SA, Gage FH *NATURE* 2001 411 6833 42-43

• This study reports the isolation and characterization of neural progenitor and stem cells from biopsies and post-mortem human brain tissues. Neural progenitor and stem cells were isolated and cultured in vitro in the presence of fibroblast growth factor, platelet derived growth factor, and the NSC factor CCg.

437418 Survival and differentiation of adult neuronal progenitor cells transplanted to the adult brain. Gage FH, Coates PW, Palmer TD, Kuhn HG, Fisher LJ, Suhonen JO, Peterson DA, Suhr ST, Ray J *PROC. NATL. ACAD. SCI. USA* 1995 92 25 11879-11883

468477 Reisolation of continuously regenerating human neural stem/progenitor AC133+ cells from transplanted mouse brain. Uchida N, He D, Buck D, Reitsma M, Masek M, Phan T, Tsukamoto A, Weissman I, Gage R *ABSTR. SOC. NEUROSCI.* 2000 26 1-2 209.9

485519 StemsCells reports promising preclinical results in NCL StemCells Inc. StemCells Inc PRESS RELEASE 2003 April 11

504337 StemCells Inc receives grant to continue development of human neural stem cells as treatment for spinal cord injuries. StemCells Inc PRESS RELEASE 2003 September 08

540074 StemCells Inc. INT BIOTECHNOL CONV and EXHIB 2004 June 6-9

559872 NIH award grants for studies involving StemCell's neural stem cells. StemCells Inc PRESS RELEASE 2004 September 20

560410 Form 10-K – StemCells Inc – for the fiscal year ended December 31, 2003. StemCells Inc FORM 10-K 2004 April 06

570810 Stem Cells and Regenerative Medicine – SRI's Fourth Annual Meeting, Commercial Implications for the Pharmaceutical and Biotech Industries, Princeton, NJ, USA. IDDB MEETING REPORT 2004 October 18-19

578153 StemCells files IND to begin Batten disease stem cell study. StemCells Inc PRESS RELEASE 2005 January 04

582789 FDA and StemCells to discuss phase I Batten's disease trial. StemCells Inc PRESS RELEASE 2005 February 01

620544 StemCells obtains license for cell processing facility. StemCells Inc PRESS RELEASE 2005 August 31

623960 Human neural stem cells differentiate and promote locomotor recovery in spinal cord-injured mice. Cummings BJ, Uchida N, Tamaki SJ, Salazar DL, Hooshmand M, Summers R, Gage FH, Anderson AJ *PROC. NATL. ACAD. SCI. USA* 2005 102 39 14069-14074

•• The release of trophic factors by grafted neural progenitor and stem cells and the interaction of these cells with the injured brain are believed to play a major role in the recovery process after transplantation. In this study, where human neural progenitor and stem cells were injected into the spinal cord after injury in mice, the locomotor recovery disappeared following treatment with diphtheria toxin (diphtheria toxin kills only human cells, not mouse cells), suggesting that the grafted cells themselves are responsible for recovery. Administration of NSC might not simply stimulate the body to produce some healing factors, but also directly contribute to repair the damages themselves.

629732 StemCells's phase I Batten disease study gets FDA approval. StemCells Inc PRESS RELEASE 2005 October 20

633798 Stem Cells and Regenerative Medicine – SRI's Fifth Annual Conference, Pittsburgh, PA, USA, 17-18 October 2005. Rippon HJ IDDB MEETING REPORT 2005 October 17-18

636193 Temporal and sequential development of human oligodendrocytes from transplanted hcns-sc. Uchida N, Capela A, Cummings B, Dohse M, Anderson A, Gage F, Back S, Tsukamoto A, Tamaki S *ABSTR. SOC. NEUROSCI.* 2005 35 718.2

636344 Quantification of engraftment and differentiation of human central nervous system stem cells (HCNS-SC) upon transplantation into the NODSCID mouse brain. Capela A, Dohse M, Jacobs Y, Reitsma M, Tamaki S, Tsukamoto A, Uchida N *ABSTR. SOC. NEUROSCI.* 2005 35 Abs 827.15

647207 Embryonic stem cells: Prospects for developmental biology and cell therapy. Wobus AM, Boheler KR *PHYSIOL. REV.* 2005 85 2 635-678

647217 AC133, a novel marker for human hematopoietic stem and progenitor cells. Yin AH, Miraglia S, Zanjani ED, Almeida-Porada G, Ogawa M, Leary AG, Olweus J, Kearney J, Buck DW *BLOOD* 1997 90 12 5002-5012

647222 Engraftment of sorted/expanded human central nervous system stem cells from fetal brain. Tamaki S, Eckert K, He D, Sutton R, Doshe M, Jain G, Tushinski R, Reitsma M, Harris B, Tsukamoto A, Gage F, Weissman I, Uchida N *J. NEUROSCI. RES.* 2002 69 6 976-986

647228 Transplanted human fetal neural stem cells survive, migrate, and differentiate in ischemic rat cerebral cortex. Kelly S, Bliss TM, Shah AK, Sun GH, Ma M, Foo WC,

Masel J, Yenari MA, Weissman IL, Uchida N, Palmer T, Steinberg GK *PROC. NATL. ACAD. SCI. USA* 2004 101 32 1183911844

647230 Adult neurogenesis and neural stem cells of the central nervous system in mammals. Taupin P, Gage FH *J. NEUROSCI. RES.* 2002 69 6 745749

647233 FGF-2-responsive neural stem cell proliferation requires CCg, a novel autocrine/paracrine cofactor. Taupin P, Ray J, Fischer WH, Suhr ST, Hakansson K, Grubb A, Gage FH *NEURON* 2000 28 2 385-397

647908 Niemann-Pick disease. Kolodny EH *CURR. OPIN. HEMATOL.* 2000 7 1 48-52

647919 Intracerebral transplantation of adult mouse neural progenitor cells into the Niemann-Pick-A mouse leads to a marked decrease in lysosomal storage pathology. Shihabuddin LS, Numan S, Huff MR, Dodge JC, Clarke J, Macauley SL, Yang W, Taksir TV, Parsons G, Passini MA, Gage FH, Stewart GR *J. NEUROSCI.* 2004 24 47 10642-10651

647946 Neural stem cells constitutively secrete neurotrophic factors and promote extensive host axonal growth after spinal cord injury. Lu P, Jones LL, Snyder EY, Tuszynski MH *EXP. NEUROL.* 2003 181 2 115-129

647954 Differentiation and tropic/trophic effects of exogenous neural precursors in the adult spinal cord. Yan J, Welsh AM, Bora SH, Snyder EY, Koliatsos VE *J. COMP. NEUROL.* 2004 480 1 101-114

647967 Neurodegenerative disease: The neuronal ceroid lipofuscinoses (Batten disease). Mitchison HM, Mole SE *CURR. OPIN. NEUROL.* 2001 14 6 795-803

647969 Disruption of PPT1 or PPT2 causes neuronal ceroid lipofuscinosis in knockout mice. Gupta P, Soyombo AA, Atashband A, Wisniewski KE, Shelton JM, Richardson JA, Hammer RE, Hofmann SL *PROC. NATL. ACAD. SCI. USA* 2001 98 24 13566-13571

654203 Human fetal CNS stem cell neurospheres survive, migrate and differentiate following transplantation into ischemic rat cerebral cortex. Kelly S, Bliss TM, Shah AK, Sun G, Puja K, Ma M, Foo W, Masel J, Shallert T, Yenari MA, Uchida N, Palmer T, Steinberg GK *ABSTR. SOC. NEUROSCI.* 2003 2003 844.9

654204 Use of human central nervous system stem cells for neurodegenerative disease and CNS injury. Tamaki S, Capela A, Dohse M, Eckert K, He D, Tushinski R, Tsukamoto A, Basu S, Belichenko P, Mobley W, Cummings B, Anderson A, Uchida N *ABSTR. SOC. NEUROSCI.* 2003 2003 656.4

654205 Transplantation of human central nervous system stem cells into a NOD –SCID mouse model of spinal cord contusion injury. Cummings BJ, Uchida N, Salazar DL, Tamaki SJ, Dohse M, Tushinski R, Tsukamoto AS, Anderson AJ *ABSTR. SOC. NEUROSCI.* 2003 2003 415.3

654206 Brain transplantation of human neural stem cells in mouse model of Batten's disease. Basu SB, Belichenko PV, Uchida N, Tamaki S, Eckert K, Udani V, Tsukamoto A, Mobley WC *ABSTR. SOC. NEUROSCI.* 2003 2003 335.7

654686 Resolution of stroke deficits following contralateral grafts of conditionally immortal neuroepithelial stem cells. Veizovic T, Beech JS, Stroemer RP, Watson WP, Hodges H *STROKE* 2001 32 4 1012-1029

654690 Immunohistochemical and electron microscopic study of invasion and differentiation in spinal cord lesion of neural stem cells grafted through cerebrospinal fluid in rat. Wu S,

Suzuki Y, Noda T, Bai H, Kitada M, Kataoka K, Nishimura Y, Ide C *J. NEUROSCI. RES.* 2002 69 6 940945

654691 Injection of adult neurospheres induces recovery in a chronic model of multiple sclerosis. Pluchino S, Quattrini A, Brambilla E, Gritti A, Salani G, Dina G, Galli R, Del Carro U, Amadio S, Bergami A, Furlan R, Comi G, Vescovi AL, Martino G *NATURE* 2003 422 6933 688-694

654808 StemCells Inc receives IRB approval from Oregon Health and Science University to begin phase I clinical trial in Batten disease; first ever clinical trial using purified human neural stem cells. StemCells Inc PRESS RELEASE 2006 March 09

655176 Survival and differentiation of neural progenitor cells derived from embryonic stem cells and transplanted into ischemic brain. Takagi Y, Nishimura M, Morizane A, Takahashi J, Nozaki K, Hayashi J, Hashimoto N *J. NEUROSURG.* 2005 103 2 304-310

Conclusion and Perspectives Adult Neural Stem Cells from Promise to Treatment: The Road Ahead

The confirmation that neurogenesis occurs in the adult brain and neural stem cells (NSCs) reside in the adult central nervous system (CNS) opens new opportunities to treat neurological diseases and injuries. To this aim, adult NSCs provide a promising model for cellular therapy. However, much remains to be done before NSC research be brought to therapy.

Stem cells are self-renewing cells that generate the various cell types of the body. In adult tissues, they contribute to homeostasis of the tissues and regeneration after injury [1]. Contrary to other adult tissues, the adult brain does not regenerate and repair itself, after injuries or diseases. It was believed that the adult CNS was composed of post-mitotic and differentiated nerve cells, born during development [2]. An underlying of this belief was that, contrary to other adult tissues, the adult brain was devoid of stem cells, hence of capacity of regeneration.

Seminal studies in the 1960s reported that neurogenesis occurs in discrete regions of the adult brain, in rodents [3, 4]. It was not until the late 80s and afterwards that it was confirmed that neurogenesis occurs in the adult brain and NSCs reside in the adult CNS in mammals, contrary to a long-held dogma [5]. Neurogenesis occurs primarily in two regions of the adult brain, the dentate gyrus of the hippocampus and the subventricular zone, along the ventricles, in various species including humans [6]. It is hypothesized that newborn neuronal cells in the adult brain originate from stem cells. NSCs are the self-renewing multipotent cells that generate neurons, astrocytes and oligodendrocytes in the nervous system. Because of their potential to generate the main phenotypes of the nervous system, they hold the potential to cure a broad range of neurological diseases and injuries.

The confirmation that neurogenesis occurs in the adult brain and NSCs reside in the adult CNS has tremendous implications for therapy. The adult CNS may be amenable to repair. To this aim, two strategies are being considered: the stimulation of endogenous neural progenitor or stem cells of the adult brain, and the transplantation of adult-derived neural progenitor and stem cells, to repair the degenerated or injured pathways.

Experimental studies reveal that new neuronal cells are generated in the diseased brain and at sites of lesions, after cerebral stokes, where they replaced some of the lost nerve cells [7-9]. Neural progenitor and stem cells have been isolated and characterized *in vitro* from the adult CNS, including from human biopsies and *post-mortem* tissues, providing a source of tissue for cellular therapy [10]. However, protocols currently devised to isolate and culture neural progenitor and stem cells yield to heterogeneous population of neural progenitor and stem cells, limiting their therapeutic potential [11]. Studies from fetal- and adult-derived neural progenitor and stem cells show that grafted cells differentiate and integrate the host tissues [12, 13]. Although cell death is still occurring and full functional recovery is not achieved, these studies reveal an attempt by the CNS to repair itself, and validate the use of adult-derived neural progenitor and stem cells for therapy.

The generation of new nerve cells at sites of degeneration or injuries, from endogenous or transplanted cells, may be insufficient to promote functional recovery. This may originate from either a low number of stimulated or grafted stem cells, or a lack of integration and differentiation, into functional cells. Stem cells reside in specialized microenvironments or "niches" that regulate their self-renewal and differentiation activities, particularly in the adult brain [14]. Hence, the microenvironment plays a key role in the therapeutic potential of adult stem cells, whether endogenous or transplanted [15]. For example, gial scar tissue at sites of degenerations and injuries is a hallmark of CNS diseases and injuries [16]. This tissue is reported to limit the regenerative potential of the CNS [17].

Hence, although adult NSCs hold the promise to treat a broad range of neurological diseases and injuries, their potential for cellular therapy may be limited by both intrinsic and extrinsic cues. Future directions will aim at unraveling the cellular and molecular mechanisms underlying the neurogenic niches, in the diseased and injured brain, and to establish homogenous population of neural progenitor or stem cells, for therapy.

Acknowledgments

Reproduced from: Taupin P. Editorial. Adult neural stem cells from promise to treatment: the road ahead. Journal of Neurodegeneration and Regeneration (2008) 1(1):7-8, with permission of Weston Medical Publishing, LLC.

References

[1] Potten CS, Loeffler M: Stem cells: attributes, cycles, spirals, pitfalls and uncertainties. Lessons for and from the crypt. *Development* 1990; 110(4): 1001-1020.

[2] Santiago Ramon y Cajal: Degeneration and Regeneration of the Nervous System. New York: Hafner (1928).

[3] Altman J, Das GD: Autoradiographic and histological evidence of postnatal hippocampal neurogenesis in rats. *J. Comp. Neurol.* 1965; 124(3): 319-336.

[4] Altman J: Autoradiographic and histological studies of postnatal neurogenesis. IV. Cell proliferation and migration in the anterior forebrain, with special reference to persisting neurogenesis in the olfactory bulb. *J. Comp. Neurol.* 1969; 137(4): 433-457.

[5] Gross CG: Neurogenesis in the adult brain: death of a dogma. *Nat. Rev. Neurosci.* 2000; 1(1): 67-73.

[6] Taupin P, Gage FH: Adult neurogenesis and neural stem cells of the central nervous system in mammals. *J. Neurosci. Res. 2002*; 69(6): 745-749.

[7] Arvidsson A, Collin T, Kirik D et al: Neuronal replacement from endogenous precursors in the adult brain after stroke. *Nat. Med.* 2002; 8(9): 963-970.

[8] Curtis MA, Penney EB, Pearson AG et al: Increased cell proliferation and neurogenesis in the adult human Huntington's disease brain. *Proc. Natl. Acad. Sci. USA* 2003; 100(15): 9023-9027.

[9] Jin K, Sun Y, Xie L et al: Directed migration of neuronal precursors into the ischemic cerebral cortex and striatum. *Mol. Cell Neurosci.* 2003; 24(1): 171-189.

[10] Palmer TD, Schwartz PH, Taupin P et al: Cell culture. Progenitor cells from human brain after death. *Nature* 2001; 411(6833): 42-43.

[11] Chin VI, Taupin P, Sanga S et al: Microfabricated platform for studying stem cell fates. *Biotechnol. Bioeng.* 2004; 88(3): 399-415.

[12] Pluchino S, Quattrini A, Brambilla E et al: Injection of adult neurospheres induces recovery in a chronic model of multiple sclerosis. *Nature* 2003; 422(6933): 688-694.

[13] Cummings BJ, Uchida N, Tamaki SJ et al: Human neural stem cells differentiate and promote locomotor recovery in spinal cord-injured mice. *Proc. Natl. Acad. Sci. USA* 2005; 102(39): 14069-14074.

[14] Moore KA, Lemischka IR: Stem cells and their niches. *Science* 2006; 311(5769): 1880-1885.

[15] Taupin P: Adult neural stem cells, neurogenic niches and cellular therapy. *Stem Cell Reviews* 2006; 2(3): 213-220.

[16] McGraw J, Hiebert GW, Steeves JD: Modulating astrogliosis after neurotrauma. *J. Neurosci. Res.* 2001; 63(2): 109-115.

[17] Ribotta MG, Menet V, Privat A: Glial scar and axonal regeneration in the CNS: lessons from GFAP and vimentin transgenic mice. *Acta Neurochir. Suppl.* 2004; 89: 87-92.

Index

B

C

D

E

F

G

H

I

N

O

P

Q

R

S

T

U

V

W

X

Y

Z